THE 'GREEN REVOLUTION' AND
ECONOMIC DEVELOPMENT

Also by Clement Tisdell

THE THEORY OF PRICE UNCERTAINTY, PRODUCTION AND PROFIT
MICROECONOMICS: The Theory of Economic Allocation
ECONOMICS OF MARKETS: An Introduction to Economic Analysis
WORKBOOK TO ACCOMPANY ECONOMICS OF MARKETS
ECONOMICS OF FIBRE MARKETS: Interdependence between Man-made
  Fibres, Wool and Cotton (*with P. McDonald*)
ECONOMICS IN OUR SOCIETY: Principles and Applications
SCIENCE AND TECHNOLOGY POLICY: Priorities of Governments
MICROECONOMIC POLICY (*with K. Hartley*)
ECONOMICS IN OUR SOCIETY: New Zealand edition (*with J. Ward*)
MICROECONOMICS OF MARKETS
WILD PIGS: Environmental Pest or Economic Resource?
ECONOMICS IN CANADIAN SOCIETY (*with B. Forster*)
WEED CONTROL ECONOMICS (*with B. Auld and K. Menz*)
TECHNOLOGICAL CHANGE, DEVELOPMENT AND THE
  ENVIRONMENT: Socio-Economic Perspectives (*co-editor with P. Maitra*)
NATURAL RESOURCES, GROWTH AND DEVELOPMENT: Economics and
  Ecology of Resource-Scarcity
ECONOMICS OF ENVIRONMENTAL CONSERVATION
DEVELOPMENT ISSUES IN SMALL ISLAND ECONOMIES (*with
  D. L. McKee*)

The 'Green Revolution' and Economic Development

The Process and its Impact in Bangladesh

Mohammad Alauddin
Lecturer in Economics
University of Queensland, Brisbane, Australia

and

Clement Tisdell
Professor of Economics
University of Queensland, Brisbane, Australia

MACMILLAN

First published 1991

Published by
MACMILLAN ACADEMIC AND PROFESSIONAL LTD
Houndmills, Basingstoke, Hampshire RG21 2XS
and London
Companies and representatives
throughout the world

Printed in Hong Kong

British Library Cataloguing in Publication Data
Alauddin, Mohammad
The 'green revolution' and economic development: the
process and its impact in Bangladesh.
1. Bangladesh. Rural regions. Economic development
I. Title   II. Tisdell, C. A. (Clement Allan)
330.9549205
ISBN 0–333–52736–4

To our families and parents

Contents

List of Tables

List of Figures

Preface

This book provides an integrated view of the effects of the 'Green Revolution' technologies in Bangladesh as they relate to growth, distribution, stability and sustainability. An attempt to draw a comprehensive picture does risk a loss of detail in any single aspect, but the wider frame of reference is necessary in order to describe the complex interrelationships between individual elements.

This book is the product of joint research which extends back over half a decade. The initial collaboration started in 1984 when the first-named author was a Visiting Fellow at the University of Newcastle, Australia. Continuing collaboration since then has resulted in a number of research publications embracing various aspects of the 'Green Revolution'. The PhD thesis of M. Alauddin in modified form constitutes parts of the present volume.

During the entire period of writing this book, we have accumulated intellectual debts to a number of scholars whose work has exerted significant influence on the development and articulation of our thoughts. They include, among others, Professors Gordon R. Conway, Romesh K. Diwan, Yujiro Hayami, Michael Lipton, Vernon W. Ruttan and Dr Peter B. R. Hazell. Discussions with Professor S. K. Saha and collaborative work with Dr Mustafa K. Mujeri formed useful ingredients of this book.

We have benefited from comments and suggestions on earlier drafts, and at various tages in our research, from Professors Kaushik Basu, Anthony Chisholm, Priyatosh Maitra and Anthony P. Thirlwall. We wish to express our deep appreciation to all of them.

The bulk of the work relating to this volume was completed at the Universities of Newcastle, Melbourne and Queensland. We wish to express our gratitude to the Departments of Economics at these universities for the use of facilities and for the congenial research atmosphere which we enjoyed during the period of writing this book. We would especially like to thank Drs Robert J. Dixon and John R. Fisher and Professor Ross A. Williams for their encouragement and support for this project. We thank Judy McCombe and her team for the drawing of the professional diagrams and Keith Povey for his excellent copy-editing of our original manuscript.

We quote from numerous publications which are acknowledged in the text. We thank the following publishers and journal editors for granting permission to include some of our previously published materials: Basil Blackwell (Chapter 2), Elsevier (Chapters 6 and 10), *Journal of Contemporary Asia* (Chapter 4), *Pakistan Development Review* (Chapter 7), *Australian Economic Papers* (Chapter 11), Frank Cass and Co. (Chapter 8), Chapman and Hall (Chapter 3), Butterworth Scientific Ltd (Chapters 12 and 13), International Association of Agricultural Economists (Chapter 8), *Developing Economies* (Chapter 9), and Routledge (Chapters 12 and 13). We also thank the editor of *American Journal of Agricultural Economics* for his permission to reproduce the Hayami-Herdt model in Chapter 6.

This book is dedicated to our families and parents. In writing this book we found it impossible to escape the impact of their profound influence and sacrifice or to appreciate fully the extent of our debt to all of them.

MOHAMMAD ALAUDDIN
CLEMENT TISDELL

1 Bangladesh's Economy, Bangladeshi Agriculture and the Issues Raised by the 'Green Revolution'

1.1 INTRODUCTION

Bangladesh has one of the lowest per capita levels of income in the world and faces formidable development problems. It is often regarded as providing an acid test for development policies, and it highlights many of the issues facing the least developed countries in the world (Faaland and Parkinson, 1976). Bangladesh's economy is overwhelmingly agricultural and this sector most reflects the country's poverty. The agricultural economy of Bangladesh provides a useful case study of peasant agriculture under conditions of abject poverty, growing landlessness, an extremely unfavourable land–man ratio and complex socio-economic and natural factors. The country's agrarian base, its inability to produce enough food to feed its growing population, the overwhelming size of its rural population and the fact that vast majority of this population lives near or below the subsistence level make sustainable growth in agriculture a *sine qua non* for its sustained economic development (Myrdal, 1968; Johnston and Mellor, 1961).

The major development influence on Bangladeshi agriculture and its economy in recent times has been the introduction of 'Green Revolution' technologies. These biochemical land-saving technologies, which are usually water or irrigation demanding, have transformed food production throughout Asia and have done much to raise rural productivity and alleviate poverty. These technologies, however, were introduced in Bangladesh later than in most other parts of Asia, including India, and appear to have been less studied in the Bangladeshi context than elsewhere. The aim of this book is to help to rectify this shortcoming.

This book concentrates on Bangladesh's economic and social experience with the introduction of Green Revolution technologies.

1

These technologies have brought about considerable socio-economic change in Bangladesh. It is useful to identify these changes and to compare and contrast Bangladesh's experiences with those in other countries. The consequences of Green Revolution technologies in Bangladesh for economic growth, the distribution of income and of resources, and for stability and sustainability of agricultural production and yields are given particular attention in this study. However, before considering in depth issues raised by the Green Revolution in Bangladesh, let us outline the broader setting in which the Green Revolution has occurred in that country.

By way of background, this chapter provides a brief description of the Bangladesh economy, its growth and structural change since the 1950s and an overview of Bangladeshi agriculture and the introduction of Green Revolution technologies. It then broadly outlines the issues to be covered in this book.

1.2 BANGLADESH ECONOMY: GENERAL CHARACTERISTICS AND GROWTH

Bangladesh, one of the world's poorest countries (World Bank, 1986), emerged as a new nation state in 1971. A recent study (BPC, 1985, p.V-4) estimates that the percentage of households having less than a prescribed minimum calorie intake (2122 kilo-calories) increased from 59 to 76 between 1975–6 and 1981–2. Bangladesh has been beset with problems inherited from years of underdevelopment and neglect (N. Islam, 1974; A.R. Khan, 1972).

Furthermore, the circumstances under which Bangladesh gained independence in 1971 created a new set of problems. Widespread destruction during the War of Liberation followed an unprecedented cyclone in 1970. In the early years of independence, a succession of droughts and floods occurred. Problems included not only the massive short-term needs of refugee rehabilitation and economic reconstruction but also long-term demands of coping with mass poverty and unemployment, malnutrition, hunger, disease and illiteracy as well as recurring natural disasters.

The economy of Bangladesh rests on a narrow base of natural resources and lacks sectoral diversification. Large-scale manufacturing contributes about 6 per cent of GDP (Gross Domestic Product) and construction and housing contributes no more than 19 per cent (BBS, 1986d, p.167). Bangladesh has no major known mineral and energy resources, with the exception of natural gas and some prospects of oil reserves.

Table 1.1 sets out selected indicators of structural change in the economy of Bangladesh during the last three decades. First, while GDP has more than doubled, so has the population. Therefore per capita GDP in the early 1980s was only marginally higher than it was three decades before. Secondly, the agricultural component of GDP has increased at a much slower rate than GDP. It has increased by more than a factor of 1.6 and overall, therefore, it has lagged behind population growth. The non-agricultural component of GDP has more than quadrupled during the same period. Thirdly, the share of non-agricultural sectors in the GDP has increased steadily over the years with a secular decline in the relative share of agriculture. Value added by the manufacturing sector rose from less than 4 per cent in the early 1950s to just over 10 per cent in recent years (Alauddin and Mujeri, 1981, p.238; BBS, 1986d, p.167). Non-industrial sectors outside the agricultural complex grew relatively rapidly. The overall slow growth in Bangladesh's GDP is due to the relatively slow growth in Bangladesh's agricultural output.

These aspects are clear from Figure 1.1. Figure 1.1a plots indices of real GDP (with the average triennium ending 1977–8 as the base) and its two components, namely agricultural and non-agricultural sectors, against time. Despite fluctuations, the overall upward movement in all the three indicators is apparent. One can also see the increasing gap between GDP and its agricultural component because of a flatter slope of the latter.

In Figure 1.1b relative shares of the agricultural and non-agricultural components have been graphed against time. A long-term declining trend in the relative share of the agricultural sector along with an increasing trend in that of the non-agricultural component can be identified. After a rapid fall in the relative share of the agricultural sector in GDP till the end of the 1970s, it appears to be stabilizing around 50 per cent.

The upward trend in GDP notwithstanding, real per capital income as illustrated in Figure 1.1c has shown little overall tendency to increase over the years. What is more disconcerting is that real GDP per head shows wide fluctuations. After showing an increasing trend in the 1960s, per capita GDP fell sharply during the early 1970s. Since then it has displayed some but not a strong tendency to increase, albeit with fluctuations.

Table 1.1   Growth of population, gross domestic product (GDP), per capita GDP and shares of agricultural and non-agricultural sectors in GDP: Bangladesh, 1950–1 to 1984–5

YEAR	POPULIN	GDP	AGRVAL	NAGRVAL	GDPPHD	AGRGDP	NAGRGDP
1950	43.3	50.648	67.267	28.775	95.713	75.471	24.529
1951	44.4	52.350	67.939	31.831	96.478	73.748	26.252
1952	45.4	53.932	70.578	32.024	97.205	74.364	25.636
1953	46.6	56.383	73.768	33.501	99.005	74.347	25.653
1954	47.7	55.124	70.862	34.409	94.562	73.050	26.950
1955	50.1	52.814	65.430	36.207	86.259	70.401	29.599
1956	51.0	58.455	74.483	37.358	93.788	72.407	27.593
1957	52.1	57.281	71.198	38.962	89.964	70.633	29.367
1958	53.1	54.081	64.206	40.755	83.339	67.464	32.536
1959	54.2	60.246	73.681	42.562	90.955	69.498	30.502
1960	55.2	63.245	77.484	44.505	93.753	69.619	30.381
1961	56.7	68.014	81.441	50.342	98.155	68.044	31.956
1962	58.2	69.227	79.104	56.226	97.330	64.934	35.066
1963	59.7	75.685	85.769	62.412	103.737	64.397	35.603
1964	61.3	76.889	86.942	63.658	102.636	64.255	35.745
1965	61.7	79.497	89.140	66.804	105.429	63.719	36.281
1966	63.5	80.275	89.451	68.199	103.444	63.321	36.679
1967	65.3	87.367	97.736	73.720	109.479	63.570	36.430
1968	67.2	89.933	96.236	81.637	109.508	60.808	39.192
1969	69.3	91.486	98.883	81.750	108.023	61.420	38.580
1970	71.0	88.264	94.438	80.139	101.724	60.800	39.200
1971	72.6	78.783	84.291	71.533	88.796	60.798	39.202
1972	74.3	79.955	84.545	73.913	88.055	60.088	39.912
1973	76.4	87.555	94.134	78.896	93.774	61.095	38.905
1974	78.0	89.306	92.251	85.429	93.688	58.700	41.300
1975	79.9	96.315	98.973	92.816	98.638	58.394	41.606
1976	81.8	98.164	95.972	101.049	98.197	55.557	44.443
1977	83.7	105.521	105.054	106.135	103.160	56.574	43.426
1978	85.6	110.865	102.753	121.543	105.979	52.667	47.333
1979	87.7	112.230	102.921	124.482	104.714	52.112	47.888
1980	89.9	119.163	108.424	133.296	108.462	51.705	48.295
1981	91.6	120.832	109.409	135.868	107.940	51.453	48.547
1982	93.6	124.993	114.484	138.823	109.271	52.048	47.952
1983	95.7	130.265	116.295	148.651	111.381	50.732	49.268
1984	99.2	135.886	118.354	158.961	112.088	49.494	50.506

Notes: 1950 means 1950–1 (July 1950 to June 1951) etc. POPULIN is population in millions, GDP, AGRVAL, NAGRVAL and GDPPHD respectively refer to indices of gross domestic product, agricultural and non-agricultural value added and GDP per capita at constant 1972–3 factor cost with the average of the triennium ending 1977–8 = 100.

Sources: Based on Alamgir and Berlage (1974, pp.161–7, 172); BBS (1979, p.340; 1984b, pp.570, 690; 1986d, p.167); EIU (1976, pp.17–18); EPBS (1969, p.120). GDP and AGRVAL data prior to 1969–70 have been converted in terms of 1972–3 prices using appropriate deflators on the basis of information contained in EIU (1976, p.17). The NAGRVAL figures are based on residuals. GDP and AGRVAL data for 1970–1 are privately obtained from the Planning Commission and are less reliable.

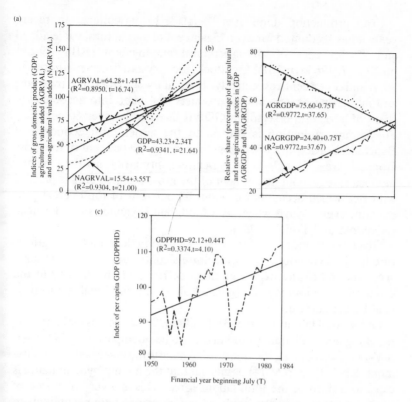

Figure 1.1   Trends in: (*a*) gross domestic product (GDP), agricultural and non-agricultural value-added (AGRVAL and NAGRVAL); (*b*) relative shares in GDP of agricultural and non-agricultural sectors (AGRGDP and NAGRGDP); (*c*) per capita GDP (GDPPHD), Bangladesh 1950–1 to 1984–5

1.3   BANGLADESHI AGRICULTURE: AN OVERVIEW

The economy of Bangladesh is characterized, *inter alia*, by the dominance of the agricultural sector (including livestock, forestry and fishery). Agriculture accounts for nearly half of the GDP, and nearly two-thirds of the labour force depends on agriculture for employment (BBS, 1985b, p.185; 1986d, p.238). Bangladesh's exports consist in the main of agricultural commodities, primarily jute and jute goods. However, remittances by Bangladeshi workers abroad (especially in the Middle East) have added significantly to the country's foreign exchange earnings in recent years.

Crop production dominates Bangladeshi agriculture. In recent years it has accounted for over 75 per cent of agricultural value added and 95 per cent of all agricultural employment (BBS, 1985b, pp.1985–7; 1986d, p.167). Livestock, forestry and fisheries contribute 10, 6 and 5 per cent respectively. If ancillary activities like transport and marketing of agricultural products are taken into account, the share of agricultural sector in GDP is likely to be over 60 per cent. Foodgrain production is central to the agricultural economy of Bangladesh. Rice and wheat together occupy 83 per cent of the gross cropped area. Bangladeshi agriculture primarily involves a rice monoculture. About 80 per cent of the gross cropped area is planted with rice which accounts for about 93 per cent of total cereal production, even though wheat as a foodgrain is gaining in importance (Alauddin and Tisdell, 1987).

Given the proportion of the total population dependent on agriculture for its livelihood, and agriculture's contribution to the national accounts and to the export earnings of Bangladesh, the key to the economic development of Bangladesh in the foreseeable future is held by agriculture.

Of the total labour force of 31.1 million, 38 per cent is estimated by the Bangladesh Planning Commission to be unemployed (BPC, 1985, p.V-14). Recognizing the difficulty of measuring disguised unemployment, BPC (1985, p.V-14) points out that the unemployment figure is close to a third of the total labour force. This is a very high rate of unemployment. Any attainable rate of industrialization is unlikely to make a significant dent in either poverty or unemployment in Bangladesh. Employment in manufacturing is constrained by imported technology which, in many cases, is not consistent with the factor endowments of the Bangladesh economy (Alauddin and Tisdell, 1986b; 1988a).[1] A partial alleviation of this problem probably lies in the expansion of the employment base within the broader context of the rural sector (BPC, 1985, p.V-2).[2]

Government aims and policies for agricultural development in Bangladesh have varied over the last four decades. The following discussion highlights the main policies and adopted objectives in agrarian planning during this period.

The most important objective of development plans in the 1950s and 1960s was to increase food production in order progressively to reduce food imports arising from the gap between demand and supply from domestic sources. This gap, this shortage of adequate local food supplies, was exacerbated by growing population pressure on the the available land, by the static nature of food production methods and of available agricultural technology. The failure to close the food gap turned a relatively strong balance of payments position in the 1950s into one of increasing deficits in the 1960s.

Faaland and Parkinson (1976, p.129) indicate that the following two-pronged strategy was adopted to deal with the situation:

(1) Large scale efforts and investments to install irrigation and drainage structures and even to control the flow of the rivers themselves;
(2) allocation of foreign exchange for large and growing quantities of foodgrain imports financed through industrialisation, import saving and export earning developments and by foreign assistance.

Although the above strategy might have been rational in conception, it was only partially implemented and often without persistence. Government efforts in advancing foodgrain production technology in the 1950s and 1960s emphasized large-scale irrigation projects of a lumpy and indivisible nature. Other technologies like chemical fertilizers, small-scale irrigation and biological innovations did not feature prominently in the Government's agricultural development strategy.[3]

Towards the later part of the 1960s, partly because foodgrain imports continued to grow and partly because there was no further possibility of bringing additional land under cultivation, a different strategy for raising food production was called for. At this stage, the new agricultural technology (designed to intensify agriculture), that is the 'Green Revolution', was promoted. This strategy commenced with the increased distribution of chemical fertilizers, followed subsequently by the introduction of more modern irrigation techniques (e.g., shallow and deep tubewells (STWs and DTWs) and low lift pumps (LLPs)) in the early 1960s. *But it was not until the later part of the 1960s*, when high-yielding varieties (HYVs) of rice and wheat were introduced, that the use of irrigation and chemical fertilizers assumed any real significance.

In the late 1960s IR-8, IR-5 and IR-20 varieties of rice continued to be introduced, initially through the direct import of seeds, and in the late 1960s and early 1970s, HYVs of wheat were introduced. Subsequently, however, the Bangladesh agricultural research system adapted and indigenously developed different varieties of rice and wheat which were multiplied and released to farmers for expanded production (Alauddin and Tisdell, 1986a).[4]

1.4   OUTLINE OF ISSUES COVERED IN THIS STUDY

During the last two decades, the traditional agriculture of many LDCs (less-developed countries) has been transformed by the introduction of new seed-fertilizer-irrigation technology. Rapid adoption of HYVs of rice and wheat has led to substantial yield gains and changes in cropping pattern in many parts of the world (Dalrymple, 1985; Herdt and Capule, 1983). While the Green Revolution is generally believed to have had favourable production consequences, its distributional impact is subject to considerable controversy (see for example, Chambers, 1984; Hayami and Ruttan, 1984; 1985; Brown, 1970; Pearse, 1980).

In addition to the productivity and distributional consequences of technological change, Conway (1986) suggests that alternative agricultural techniques should be assessed on the basis of their influences on (1) the stability of yields and incomes (stability); and (2) the *sustainability* of production and yield (sustainability). Recent studies (Hazell, 1982; 1985; 1986; Mehra, 1981) indicate that production and yield of foodgrains may have become more unstable in the period following the introduction of new agricultural technology in a number of countries (e.g. India). Recently the question of sustainability of food production, given technological change, has received renewed attention (Douglass, 1984).

The focus has shifted to some extent to considering the sustainability of the ecosystems and environmental factors on which agriculture depends. This is clear, for example, from the report commissioned by the General Assembly of the United Nations (WCED, 1987, especially ch. 5). Taking this into consideration, scientists and agricultural economists are far from unanimous about the extent to which high levels of production based on such modern technology can be sustained (Conway, 1986; Schultz, 1974; Tisdell, 1988). The use of agro-chemicals, it is argued, can reduce species diversity, upset ecological balance and stimulate the development of pest populations. Concerns are also expressed that the Green Revolution rests on a narrow range of genetic materials (see for example, Biggs and Clay, 1981).

While the rate of introduction of the new technology to Bangladeshi agriculture has been slower than in other countries of the Indian subcontinent, this technology has substantially transformed agricultural production in Bangladesh (for further details see Alauddin and Tisdell, 1986c; 1987). Against this background, the present volume examines the impact of Green Revolution technologies in Bangladesh on: (1) growth of output and yield, (2) distribution of income and resources and rural poverty, (3) stability of production and yield, and (4) sustainability of production and yield. Economic effects are examined in terms of multiple attributes rather than a single attribute. The study is multidimensional rather than unidimensional, unlike many cost–benefit analyses.

Within the Bangladeshi context, we endeavour to answer questions such as the following in relation to the Green Revolution:

1. How successful has the Green Revolution been in augmenting growth rates in output and yield of agricultural products?
2. Has there been any change in the supply base of Bangladeshi agriculture, that is, the composition of agricultural products supplied as a result of the Green Revolution?
3. How have agricultural innovations contributed to increased land productivity?
4. Are there different regional patterns in the agricultural growth process? If so, how can they be explained and what implications can be drawn from them about the effect of the new technology?
5. What are the consumer and producer welfare effects of the growth in crop output resulting from the adoption of the new technology?
6. What are the patterns and determinants of farm-level adoption and diffusion of the new technology?
7. What implications do Green Revolution technologies at the aggregate and farm levels have for income distribution and rural poverty?
8. Has the Green Revolution been a source of increased production and yield variability, and has it led to greater crop diversification?
9. What are Bangladesh's prospects for sustainable food production, and what obstacles is it likely to face in this regard?

The first four questions relate to the growth theme of the book. The next three questions are pertinent to the distribution theme, and the last two questions relate to the stability and sustainability themes. Individual chapters addressing these issues are organized as follows:

Chapter 2 outlines and investigates growth and changes in the Bangladeshi crop sector over the last four decades, using aggregate time series data. The later part of Chapter 2 employs and critically appraises non-production function decomposition methods (Minhas and Vaidyanathan, 1965; Venegas and Ruttan, 1964; Wennergren *et al.*, 1984) to identify apparent sources of crop output growth by component elements. This chapter provides a more rigorous and in-depth analysis of the underlying sources of Bangladeshi agricultural growth than previous studies employing these methods (Mahabub Hossain, 1984; Pray, 1979).

Chapter 3 assesses quantitatively the impact of biological technology in increasing the productivity of cultivated land following the Green Revolution. The model of Diwan and Kallianpur (1985) is applied to estimate the contribution of biological technology in increasing Bangladeshi foodgrain productivity per hectare of cultivated land. Our findings for Bangladesh are then compared with those of Diwan-Kallianpur (D-K) for India. Based on the conclusions reached via reinterpretation of the D-K model and supported by other evidence, it is contended that the D-K approach is subject to severe limitations and that it can give a misleading view of the contribution to production of the Green Revolution technology.

Chapter 4 provides new information about growth rates of foodgrain production in Bangladesh and about yields by districts, that is, in a more disaggregated way than in earlier chapters. Using the same framework of estimating growth rates as that used in Chapter 2, Chapter 4 examines the district-level foodgrain production and yield performance for the periods prior to and following the Green Revolution and provides estimates of district-level growth rates in foodgrain production and yield. In Chapter 4 the estimated growth rates are employed to identify regular regional patterns and whether growth rates are greater in the period following the Green Revolution. Furthermore, factors underlying different regional patterns are identified. Particular attention is paid to the rate of diffusion of important components of the new agricultural technology by regions. A cross-sectional approach is used to isolate important elements in agricultural growth. This chapter extends the analysis of district-level foodgrain production and yield of some previous studies (e.g. Alauddin and Mujeri, 1986a; 1986b).

Chapter 5 considers possible consumer welfare implications of the differential pattern of growth in the output of crops. Particular attention is given to whether the average Bangladeshi diet is less varied and balanced and *a priori* less nutritious than prior to the Green Revolution. Price and income trends and Hicksian indifference curve analysis are employed to discuss the welfare implications.

Chapter 6 specifically deals with the distributional consequences of the Green Revolution technology. To what extent have technological changes, in fact, benefited consumers and producers? Which group of producers has benefited more from technological change? Has the Green Revolution accentuated income inequality? These questions are considered by applying a model employed by Hayami and Herdt (1977).

The Hayami–Herdt (H-H) model is applied to Bangladeshi data and the findings are compared to those of Hayami and Herdt (1977). Shortcomings are outlined, and significant limitations are illustrated by data from Bangladesh. It is contended that the H-H model fails to provide a realistic assessment of the income distributional con-sequences of the Green Revolution. Furthermore, it is argued that inappropriate economic inferences can be drawn from the H-H model.

Chapter 7 examines the pattern of adoption of HYV technology in Bangladesh, employing farm-level data from two Bangladeshi vil-lages. The objective of this chapter is to provide recent evidence to test existing theories and hypotheses about adoption and diffusion patterns of HYVs. Previous studies investigating patterns and deter-minants of HYV adoption in Bangladesh (e.g. I. Ahmed, 1981; Asaduzzaman, 1979; Mahabub Hossain, 1988; Jones, 1984; Atiqur Rahman, 1981) either lack analytical rigour, take a partial view or employ data of the early 1970s.

While Chapter 6 focuses on the distribution of income generated through the market system, Chapter 8 addresses broader distri-butional issues embracing technological change, resource distribution and rural poverty. In particular, it is emphasised that while the market system plays a significant part in determining the income of the rural poor, a proportion of their income is determined outside the market system, and in this respect their access to natural resources is of significance for their real level of income. To the extent that new agricultural technology has deprived the rural poor of their easy access to natural resources, it has reduced their incomes.

Given technological and other changes in Bangladeshi agriculture and the resulting increase in overall food production and yield per hectare, it is worth while asking questions of distributional relevance: How have these changes affected the distribution of land and access to natural resources? Have real wages of agricultural labour in-creased in recent years? Have the opportunities for supplementary incomes been reduced with population pressure and greater pen-etration of the new technology? Do farmers feel that their economic situation is more vulnerable to poverty risk than before? Does this vulnerability, if any, vary across farms? Have the conditions of the rural poor consisting of landless labourers and marginal farmers significantly improved? Have the distributions of income and ownership of assets become more unequal in recent years?.

Chapter 8 provides some answers to the questions. Amongst other things, Chapter 8 also considers whether increased privatization of lands formerly having common access, e.g. animal watering points, grazing lands, etc., has led to the exclusion of the rural poor with adverse consequences for poverty and income security. Previous studies of rural poverty in Bangladesh (e.g. A.R Khan, 1984) have not incorporated the non-exchange component of incomes of the rural poor and have not examined implications of declining access to land and other resources for income security.

In the light of recent developments in the input as well as in the output side of Bangladeshi foodgrain production, Chapter 9 analyses data on the variability of Bangladeshi foodgrain production and considers the possible role of the new technology in moderating or accentuating fluctuations in production and yield. In the discussion, the main focus is on the aggregate supply of foodgrains (rice and wheat).

Chapter 9 also briefly reviews the relevant literature and contends that the methodologies adopted by the authors of some recent studies (e.g. Hazell, 1982; Mehra, 1981) suffer from serious shortcomings, can produce misleading conclusions and be a source of faulty policy advice. An alternative methodology is then applied in order to determine trends in the variability of Bangladeshi foodgrain production and yield over time. Factors which may explain the underlying trends in the variability of foodgrain production are discussed.

In Chapter 10, evidence from Bangladesh about changes in variability of production and yield of foodgrains prior to and following the introduction of the HYVs of cereals and associated techniques is outlined and investigated on a regional basis, that is, at a more disaggregated level than earlier. Some pertinent questions are asked and addressed: To what extent have technological changes led to alterations in the variability of overall foodgrain production and yield? Is there any significant difference in production and yield variability between traditional and modern foodgrain varieties? Has irrigation had a stabilizing impact on production and yield fluctuations?

These questions are addressed using Bangladeshi national and regional data with special emphasis on the latter. First of all, the methodology employed for analysing the data is specified. Trends in foodgrain production and yield variability for Bangladesh as a whole, as well as in the main regions (districts) of Bangladesh are then discussed. District (regional) data are also used to examine how variability of foodgrain production and yield have altered with the introduction of new agricultural technology. In addition, this chapter investigates whether variability in foodgrain production and yield variability are systematically related to the rate of diffusion of new agricultural techniques.

Chapter 11 extends earlier research reported in Chapters 9 and 10 by considering the impact of the introduction of HYVs in Bangladesh on the lower semi-variance of foodgrain yields, on its lower semi-variation, and the probability of disaster yield levels, using a modification of Cherbychev's inequality. The trend in these characteristics is downwards and this is highlighted by a favourable shift in the empirically derived mean yield/risk efficiency locus. A number of factors may explain this shift. Using portfolio diversification theory and drawing on survey evidence, particular attention is given to greater crop diversification as a *possible* explanation, but it is suggested that it is not important in practice. Indeed, in the long-term, reduced genetic diversity seems likely and may pose new risks.

Chapter 12 provides new estimates and interpretations of trends in Bangladeshi foodgrain production and yield. The evidence about past growth of foodgrain production is reviewed and a somewhat different view to that of Boyce (1985, 1987a) is provided. Projections for Bangladeshi foodgrain production and yield are given on a different basis to that normally used in the literature, and particular account is taken of whether growth of foodgrain production and yield can be sustained. The food security situation of Bangladesh is also discussed by examining trends in the availability of foodgrain from domestic production and trends in the import intensity of foodgrains.

Sustainability issues touched on in Chapters 11 and 12 are focused on in Chapter 13. This considers the prospects for sustaining Bangladesh's foodgrain production using the new agricultural technology. Consideration is given to whether Bangladesh's food production is sustainable given: (1) its increasing dependence on new ('modern') agricultural technology; (2) its reliance on foreign technology; and (3) its dependence on imports of inputs required for 'modern agriculture'.

The sustainability issue has not been previously investigated for Bangladeshi agriculture but Bangladesh's situation is not unique in the world. Its food sustainability problem has parallels with those of many developing countries (see for example, Chisholm and Tyers, 1982) and this makes its case of additional interest.

Chapter 14 provides an overview of the results of this study and suggests an agenda for additional research.

Thus within the framework of this book the implications are considered (in the Bangladeshi context) of the Green Revolution for: (1) growth of agricultural output and yield per hectare; (2) distribution of rural income and resources, and occurrence of rural poverty; (3) instability of crop production and yield; and (4) sustainability of crop production. New hypotheses and empirical evidence about the impact of the Green Revolution on economic development are presented. But the discussion is not merely confined to an assessment of the impact of the Green Revolution. For example, patterns and the dynamics of adoption and the diffusion of the new agricultural technology and elements influencing these are identified. The value of our results are strengthened because they are compared with those of other authors and across countries.

2 New Technology, Growth Rates and Sources of Increased Agricultural Output Growth in Bangladesh

2.1 INTRODUCTION

Crop production is central to the agricultural economy of Bangladesh and therefore, in considering increased agricultural output in Bangladesh, we concentrate on crop produciton. The critical importance of the crop sector to Bangladeshi agriculture apart, it is analysed because of the availability of relatively more reliable time series data than are available for livestock, forestry and fisheries. Furthermore, the technological changes generally identified with the Green Revolution are confined to crop production and primarily to rice and wheat and to some extent to jute and sugar cane. The introduction of the new technology in Bangladesh, as in many LDCs, has substantially altered the nature of agricultural production (Dalrymple, 1985; Herdt and Capule, 1983; Staub and Blase, 1974; Alauddin and Mujeri, 1986a, 1986b).

This chapter investigates growth and changes in the crop sector of Bangladesh over the last four decades and in so doing attempts to avoid limitations of previous studies of this subject. It (1) uses a variant of the kinked exponential model; (2) covers a greater number of crops and commodity groups; and (3) employs more recent data. Two sets of estimates are made:

1. For the whole period 1947–8 to 1984–5, dividing it into three sub-periods and thereby two kinks into the exponential growth function. In this exercise, 21 crops are considered of which 13 are non-cereals. These are divided into four groups, namely Cereals, Cash Crops, Oilseeds and Pulses.
2. More detailed and comprehensive estimates for the period 1967–8

to 1984–5 are made. This period is divided into two sub-periods and a single kink is thereby introduced. Compared to the earlier analysis, the number of crops is extended to 36. The number of crop groupings is increased by the addition of the groups Fruits, Vegetables, and Spices and the pre-existing Pulses group is expanded by the addition of peas and *khesari*.

The later part of this chapter employs and critically appraises non-production function decomposition methods (Minhas and Vaidyanathan, 1965; Venegas and Ruttan, 1964; Wennergren *et al.*, 1984) to identify apparent sources of crop output growth by component elements. This chapter provides a more rigorous and in-depth analysis of the underlying sources of growth than some previous studies employing these methods (M. Hossain, 1984; Pray, 1979).

2.2   METHODOLOGICAL ISSUES OF GROWTH RATE ESTIMATION

A number of recent studies (Alauddin and Mujeri, 1986a; M. Hossain, 1984; Boyce, 1985) have examined the growth of Bangladeshi crop output. Alauddin and Mujeri (1986a) derived growth rates in aggregate crop output (involving 13 crops) between 1967–9 (average for the years 1967–8, 1968–9 and 1969–70) and 1978–80 (average for the years 1978–9, 1979–80 and 1980–1). The estimated output growth rate is a function of only two end-points and is quite independent of the actual process underlying it. The growth rate derived on the basis of such end-points may be of limited predictive value (Rudra, 1970). M. Hossain (1984) derives growth rates of aggregate crop production on the basis of 15 major crops. However, as discussed later in this chapter, Hossain's approach is arbitrary in the choice of both functional form and time cut-off points. Boyce's (1985) estimates suffer from similar limitations (Alauddin and Tisdell, 1987). In a subsequent paper, however, Boyce (1986; see also Boyce, 1987b) uses a kinked exponential model which is distinctly superior to the methods of Hossain (1984) and Boyce (1985).

Two aspects of the analytical framework warrant discussion: (1) choice of sub-periods, and (2) functional form. The choice of appropriate sub-periods is determined by *a priori* considerations such as known turning points in the history of the time series. The turning points themselves depend on a number of factors, including introduction of innovations, political changes, and natural factors like a severe drought or a flood.

In recent studies of Bangladeshi agriculture, a variety of sub-periods have been chosen for consideration. Boyce (1985; 1986), for instance, in estimating growth rates of agricultural production, has divided the 1949–80 (1949–50 to 1980–1) period into two sub-periods of equal length, the dividing line being the year 1964–5 which marks the end of the first sub-period. Boyce's choice of the cut-off point seems to have been made to compare growth rates during the periods of traditional and new agricultural technologies (Boyce, 1986, p.388). However, the rationale of choosing 1964–5 as the dividing line between the two technological phases is not clear. If one were to assume that the central concern of the Boyce thesis is the role of technological change accompanying the Green Revolution in Bangladesh, the choice of one of the years between 1967–8 and 1969–70 as the cut-off point would have been more logical.

M. Hossain (1984, p.32) divides the 1949–50 to 1983–4 periods into three sub-periods: (1) 1949–50 to 1957–8; (2) 1958–9 to 1970–1 and (3) 1970–1 to 1983–4. Hossain's choice seems to have been based on political considerations. Hossain defines his first sub-period as 'the period of political instability before the military take-over in 1958', the second as the 'Ayub regime'[1] and the third as 'the independence of Bangladesh' in 1971. Thus in his choice of sub-periods Hossain does not seem to have paid much attention to the introduction of HYVs in the later part of the 1960s, even though his third phase is entirely within the period of the new technology.

The choice of sub-periods in the present study is based on a combination of technological, natural and political factors. The entire period 1947–84 can be broadly divided into two sub-periods: 1947–66 (1947–8 to 1966–7) and 1967–84 (1967–8 to 1984–5). The first is identified with the traditional and the second with the new technology. The earlier part of the second sub-period is marked by political and natural factors such as the War of Liberation (1970–1 and 1971–2) and droughts and floods in 1972–3, 1973–4 and 1974–5 which seriously disrupted agricultural production. It was not until 1975–6 that foodgrain production surpassed that of 1969–70 which was the pre-Liberation peak. Furthermore, the year 1975–6 saw the introduction of HYV wheat cultivation on a massive scale. Considering all these factors, the 1967–84 sub-period can be subdivided into two intervals: 1967–74 (1967–8 to 1974–5) and 1975–84 (1975–6 to 1984–5). So in effect the time series used here consists of three sub-periods: 1947–66, 1967–74 and 1975–84.

Apart from statistical considerations, e.g. explanatory power of the estimated equation, statistical significance of the coefficients, the choice of functional forms for estimating growth rates depends on a number of factors including the objective of investigation. For example, if the objective is to see if the growth rate is declining, increasing or remaining more or less stationary, then a Gompertz curve may be fitted (Reddy, 1978; Rudra, 1970). If, on the other hand, the objective is to investigate whether the trend is slowing down or the variable in question is approaching an upper limit, a logistic function or a modified exponential function may be fitted (Chakrabarti, 1982). For forecasting purposes, a log quadratic function, which assumes constant acceleration or deceleration may be appropriate. If, however, the objective is just simply to know the constant absolute (or relative) amount by which a variable changes with respect to time, then a linear (or an exponential) function seems more appropriate. But the choice of the linear or exponential function itself imposes *a priori* restrictions on the growth path of the variable which may in reality differ from the one stipulated by estimated functions.

One objective of this chapter is to compare growth rates of various crops within and between sub-periods. The conventional way is to fit trend exponential lines by ordinary least squares (OLS) technique to each segment of a time series (see, for example, World Bank, 1986). Alternatively growth rates for, say, two sub-periods can be estimated by fitting a single equation involving a dummy variable:

$$1nY_t = \alpha_1 D_1 + \alpha_2 D_2 + (\beta_1 D_1 + \beta_2 D_2)t + e_t \qquad (2.1)$$

where $t = 0,\ldots, n$ and there is a trend break at point k. D_i is a dummy variable which assumes the value 1 for the ith sub-period or 0 otherwise.

Boyce (1986) analytically argues and empirically demonstrates with Bangladeshi and West Bengal data that such a technique is likely to introduce a discontinuity bias and lead to anomalous results. For instance, a discontinuous function, Boyce (p.389) found that the Indian state of West Bengal experienced more rapid agricultural growth (1.67 per cent and 2.73 per cent) than Bangladesh (1.27 per cent and 2.18 per cent) during both sub-periods, 1949–64 and 1965–80. When exponential trends were fitted to the entire series (1949–80), Bangladesh was found to have experienced a higher rate of growth than West Bengal (2.03 per cent as against 1.75 per cent). Boyce indicates how 'unusual' deviations in the neighbourhood of break point can bias the growth results.

In order to eliminate this bias between the two trend lines, Boyce, following Poirier (1976), proposes a linear restriction such that the two lines intersect at the break point k:

$$\alpha_1 + \beta_1 k = \alpha_2 + \beta_2 k \tag{2.2}$$

Noting that $\alpha_1 D_1 + \alpha_2 D_2 = \alpha_1$ and eliminating α_2 from Equation (2.1) the restricted form is:

$$1nY_t = \alpha_1 + \beta_1(D_1 t + D_2 k) + \beta_2(D_2 t - D_2 k) + e_t \tag{2.3}$$

A straightforward application of the OLS technique yields estimates of β_1 and β_2 which give exponential growth rates corresponding to the two sub-periods. Thus the relation $\beta_1 \neq \beta_2$ implies the presence of a kink at the point k between the two sub-periods.

In case of a time series consisting of two kinks, say, k_1 and k_2, i.e. three sub-periods, the discontinuous model is:

$$1nY_t = \alpha_1 D_1 + \alpha_2 D_2 + \alpha_3 D_3 + (\beta_1 D_1 + \beta_2 D_2 + \beta_3 D_3)t + e_t \tag{2.4}$$

β_1, β_2 and β_3 are estimated growth rates which correspond to those obtained from separately fitting exponential trend to the relevant sub-period data. The presence of two kinks, k_1 and k_2 implies the following linear restrictions of the sub-period trend lines intersecting at k_1 and k_2:

$$\alpha_1 + \beta_1 k_1 = \alpha2 + \beta_2 k_2 \text{ and } \alpha_2 + \beta_2 k_2 = \alpha_3 + \beta_3 k_3 \tag{2.5}$$

Eliminating α_2 and α_3 from Equation (2.4), the two-kink model takes the following form:[2]

$$1nY_t = \alpha_1 + \beta_1(D_1 t + D_2 k_1 + D_3 k_1) + \beta_2(D_2 t - D_2 k_1 - D_3 k_1$$
$$+ D_3 k_2) + \beta_3(D_3 t - D_3 k_2) + e_t \tag{2.6}$$

A kinked exponential model seems to have the distinct advantage over a discontinuous model in that it makes use of the entire set of observations and is, therefore, likely to eliminate discontinuity bias.

Therefore, in comparing growth rates between sub-periods for the same crop or within periods for different crops, a kinked exponential model is applied in preference to a discontinuous function involving time dummies. Ideally, the standard dummy variable approach does require a discrete change in circumstances at the break point. While this may occur, in practice change is more gradual in most instances, even in the case of the introduction of Green Revolution technologies. When this is so it may tend to make the kinked method more appropriate. One needs to bear in mind that the main limitation of an exponential trend is that it yields a constant growth rate within a particular time series (or a segment thereof).

It must be emphasized, however, that 'an inevitable element of arbitrariness remains in any growth estimation procedure: the choice of end points of the time series, the demarcation of subperiods, and the inclusion and exclusion of exceptional observations may affect results. Moreover, for forecasting purposes an alternative functional form. . .might be more appropriate' (Boyce 1986, p.390). The discussion surrounding these issues warrants further investigation. So for the time being we proceed with the kinked exponential model as outlined above.

2.3 PATTERN OF CROP OUTPUT GROWTH: 1947–8 TO 1984–5

On the basis of the methodological framework discussed in Section 2.2, the objective of this section is to estimate trends in output of major commodity groups and some of the important crops within these. This will enable us to determine whether the rates of growth of individual crops and commodity groups have changed from one sub-period to another. Furthermore, it is of interest to identify any unevenness in the growth of various crops and any discernible pattern in the output-mix over the years.

The present analysis is based on official data, the reliability and accuracy of which is less than perfect (see, for example, Boyce 1985; Pray 1980). However, while revised data have been suggested by Boyce and by Pray, it is doubtful if the revised series based on these data is decisively superior. Alauddin and Tisdell (1987, p.325; see also Alauddin 1988) argue that this may not be so (cf. Boyce 1987a). Despite significant limitations of the official data, there is no other comprehensive source of agricultural statistics in Bangladesh. In these circumstances, we have no choice but to use them while acknowledging their shortcomings.

Even though a large number of crops are grown in Bangladesh, three crops – rice, wheat and jute – account for the bulk of total cropped area. The analysis of this section concentrates on four broad commodity groups as follows:

Cereals: Six varieties of rice: *aus* local, *aus* HYV, *aman* local, *aman* HYV, *boro* local and *boro* HYV; and two varieties of wheat: wheat local and wheat HYV.

Cash Crops: Jute, sugar cane, tea and tobacco.

Oilseeds: Rape and mustard, sesame seeds, linseed, groundnut and coconut.

Pulses: Lentils, gram, *mashkalai* and *moong*.

The 21 crops listed above account for more than 94 per cent of gross cropped area in Bangladesh in recent years. To facilitate aggregation, output has been measured in value terms using the 1981–3 harvest prices of agricultural commodities (average for the years 1981–2, 1982–3 and 1983–4). The average prices have been derived using respective production figures as weights.

As the time series has been broken at two points to constitute three sub-periods, Equation (2.6) is employed to estimate rates of crop output growth. The results of the exercise are set out in Table 2.1. The estimated growth rates are derived using index numbers of the relevant variables with the average of the triennium ending 1977–8 as the base.

It can be clearly seen that various crops have experienced varying rates of growth, both inter-temporally and intra-temporally. In Sub-period 1 all the crops or commodity groups except pulses registered positive rates of growth. The growth rates in cereals in all the sub-periods are closer to those of rice. But it is closest in the 1947–66 period and furthest in the third sub-period, suggesting the relative importance of wheat, the output of which has increased at a phenomenally high rate of nearly 24 per cent per annum in the latest sub-period. Cash crops, particularly jute, suffered greatest output decline in the second sub-period. Even though all the non-cereal commodity groups have experienced positive rates of output growth in the 1975–84 period, the statistical significance of the estimated growth rates are not high enough to be confident about their quality. Despite this, it seems that the output of non-cereal crops, while

Table 2.1 Estimated growth rates in output and yield of major crops for different periods, using a two-kink exponential model: Bangladesh, 1947–8 to 1984–5

Crop	Intercept	Growth rate (per cent) 1947–66	1967–74	1975–84	R^2	F-Ratio
(a) Output						
Cereals	3.96459	2.17	1.98	3.00	0.9028	105.239[a]
t-value		8.40[a]	3.64[a]	3.43[a]		
Rice	3.97779	2.16	1.81	2.33	0.8921	93.679[a]
t-value		8.40[a]	3.34[a]	2.68[a]		
Wheat	1.79296	6.10	18.70	23.62	0.9688	352.124[a]
t-value		7.62[a]	11.11[a]	8.73[a]		
Cash crop	4.36298	1.80	−1.38	1.49	0.4571	9.543[a]
t-value		4.81[a]	1.76[b]	1.18[d]		
Jute	4.68852	1.04	−3.05	0.89	0.1826	2.531[c]
t-value		1.81[b]	2.52[a]	0.46[d]		
Pulses	4.94783	−1.62	−0.075	−1.84	0.4710	10.090[a]
t-value		3.59[a]	0.08[d]	1.21[d]		
Oilseeds	4.13319	2.27	−0.02	0.49	0.6695	22.956[a]
t-value		6.21[a]	0.03[d]	0.40[d]		
Non-cereals	4.39691	1.49	−1.05	1.06	0.4511	9.316[a]
t-value		4.74[a]	1.59[c]	1.00[d]		
All crops	4.03762	2.03	1.51	2.75	0.9053	108.375[a]
t-value		9.02[a]	3.18[a]	3.60[a]		
(b) Yield (per gross cropped hectare)						
Cereals	4.25305	1.17	1.00	2.05	0.8304	55.488[a]
t-value		5.95[a]	2.42[a]	3.07[a]		
Rice	4.25389	1.19	0.97	1.95	0.8271	54.204[a]
t-value		6.01[a]	2.34[a]	2.92[a]		
Wheat	3.45196	1.66	8.28	5.44	0.9236	137.010[a]
t-value		3.71[a]	8.79[a]	3.60[a]		
Cash crops	4.34345	−0.048	3.13	−0.11	0.4223	8.283[a]
t-value		0.11[d]	3.28[a]	0.07[d]		
Jute	4.75917	−0.55	−0.95	2.39	0.1713	2.342[c]
t-value		1.29[c]	1.06[d]	1.66[b]		
Pulses	4.58013	0.36	−0.74	−0.32	0.1541	2.064[d]
t-value		1.80[b]	1.75[b]	0.48[d]		
Oilseeds	4.22184	1.71	0.63	0.87	0.7532	34.585[a]
t-value		6.41[a]	1.12[d]	0.96[d]		
Non-cereals	4.34248	0.54	1.77	0.46	0.6064	17.462[a]
t-value		1.99[b]	3.13[a]	0.51[d]		
All crops	4.28168	1.05	0.99	1.84	0.8585	68.763[a]
t-value		6.46[a]	2.88[a]	3.33[a]		

Table 2.1   *continued*

Crop	Intercept	Growth rate (per cent) 1947–66	1967–74	1975–84	R^2	F-Ratio
(c) Yield (per net cropped hectare)						
Rice	4.11558	1.64	1.58	2.89	0.8723	77.450[a]
t-value		6.79[a]	3.11[a]	3.54[a]		
Cereals	4.11463	1.62	1.60	2.98	0.8740	78.618[a]
t-value		6.73[a]	3.17[a]	3.66[a]		
All crops	4.09497	1.50	1.59	2.77	0.8893	91.010[a]
t-value		7.09[a]	3.57[a]	3.88[a]		

[a] Significant at 1 per cent level; [b] Significant at 5 per cent level; [c] Significant at 10 per cent level; [d] Not significant.

Source: Based on data from BBS (1976, pp.1–2, 4–7, 12–21, 26–9, 66–9, 74–87, 90–1, 104–19, 122–4, 126–9, 144–52, 154–7, 159, 166–9, 176, 198–205, 218, 221–7, 258–9, 261, 272–7; 1979, pp.160, 166–71, 182–4, 190–2, 197, 202–5, 207, 368; 1980a, pp.20–5, 30–1, 33–4, 36–7; 1982, pp.230, 232, 235–8, 240–1, 249; 1984a, pp.31, 39, 42; 1984b, pp.244, 249–51, 256–65, 267, 268, 270, 273, 274, 276, 278–9, 281, 283; 1985a, pp.24–6, 56–8, 157–9, 203–5, 221–3, 236, 383–5, 387–9, 399–401, 485–6, 792–806; 1985b, pp.258, 303–28; 1985c, pp.426–35; 1986a, pp.39, 47–66; 1986d, pp.25, 31–48; BRRI (1977, pp.89); EPBS (1969, pp.40–1); World Bank (1982, Tables 2.5, 2.20).

differing among themselves, has grown at a much slower pace than that of cereals. Overall output-mix seems to be consisting of a relatively greater percentage of cereals and a smaller percentage of non-cereals. Thus a discernible pattern away from non-cereals and in favour of cereals over the years can be identified. This is illustrated in Figure 2.1.

In analysing rates of growth of yields of various crops, two types of yields have been distinguished: yield per gross cropped hectare (i.e. including multiple cropping) and yield per net cropped hectare (i.e. gross cropped hectare deflated by the intensity of cropping index). Growth of yields per gross cropped area relative to output for cereals are higher during the later sub-periods. Non-cereals have in general experienced higher positive yield growth compared to their output in later sub-periods, even though one needs to be wary about the quality of estimates. The relatively high rates of growth of yields of non-cereal crops in the second sub-period compared to their lower or even negative output growth in the same period indicate significant

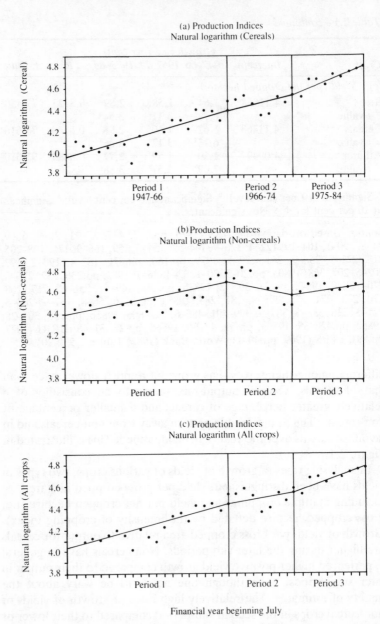

Figure 2.1   Differential pattern of crop output growth: Cereals and non-cereals in Bangladesh, 1947–8 to 1984–5

changes in the cropping pattern away from these crops following higher increases in cereal prices in the early 1970s. Jute yield grew at a much higher rate in the 1975–84 sub-period, following the introduction of the Intensive Jute Cultivation Scheme (IJCS). The growth rates of yields per net cropped hectare have been set out in part (c) of Table 2.1. The impact of the new agricultural technology comes into sharper focus as the growth rates in net yields edge closer to the respective output growth rates in the later sub-periods. Overall crop output growth is significantly higher in the third sub-period and exceeds the rate of growth of population. Overall the output and productivity growth in non-cereals commodity groups seems to lag far behind those in the cereal sub-sector as well as population growth.

The analysis in this section is based on 21 crops (13 of these being non-cereals). To identify any emerging pattern of trend in favour of or away from any commodity group one needs to include a greater number of crops. Time series data for most crops not considered in the preceding analysis because of data constraints are available from the late 1960s onwards. Therefore the next section discusses the consequences of the new technology for overall input and output mix of agricultural crops by extending the analysis to seven commodity groups consisting of 36 crops for the period 1967–8 to 1984–5.

2.4 INDICATORS OF TECHNOLOGICAL CHANGE IN BANGLADESH: GROWTH AND CHANGE IN CROP OUTPUT IN THE PERIOD OF THE NEW TECHNOLOGY

This section examines growth rates of crop output in Bangladesh since 1967–8. The analysis is carried out for seven broad groups consisting of 36 crops. The commodity groups designated Cereals, Cash crops and Oilseeds consist of the same crops as in Section 3. Two more crops, peas and *khesari*, have been added to the list of pulses, and three more commodity groups consisting of 13 crops have been included. The 36 crops constitute 97 per cent of the gross cropped area. The redefined commodity group, Pulses, and the three additional ones are as follows:

Pulses: Lentils, gram, *mashkalai*, *moong*, peas and *khesari*.
Fruits: Mango, banana, pineapple and jack fruit.
Vegetables: Potato, sweet potato, tomato, cobbage and cauliflower.
Spices: Chilli, onion, garlic and turmeric.

Table 2.2   Selected indicators of technical change in Bangladesh's agriculture: 1967–9 to 1982–4

Indicator	1967–9[a]	1982–4[b]	Per cent change
1 Total cropped area ('000 ha)	12 871	13 231	2.80
2 Net cultivated area ('000 ha)	8786	8635	−1.73
3 Area under rice cultivation ('000 ha)	9987	10 452	4.72
4 Rice area as percentage of total cropped area	77.6	79.1	1.93
5 Area under HYV rice ('000 ha)	162	2678	1553.37
6 HYV area as percentage of total rice area	1.6	25.6	1500.00
7 Area under wheat ('000 ha)	105	574	446.48
8 Area under HYV wheat ('000 ha)	6.4	554	8665.25
9 HYV wheat area as percentage of total wheat area	6.1	96.5	1481.9
10 Total area irrigated ('000 ha)	1060*	1947	83.66
11 Irrigated area as percentage of total cropped area	8.0*	14.71	83.88
12 Irrigated foodgrain area as percentage of total irrigated area	85.8*	89.62	4.45
13 Irrigation by methods Modern (%)[c]	31.5*	76.70	143.49
Traditional	68.5*	23.30	−65.99
14 Total fertilizer used ('000 m/ton nutrient)	113.1	553.3	391.02
15 Fertilizer use (kg/ha)	8.8	42.0	377.27
16 Cropping intensity	146.5	153.2	4.57
17 Foodgrain production ('000 m/tons) Rice	11 669	14 449	23.82
Wheat	97	1257	1196.44
Total	11 766	15 706	33.49
18 Foodgrain yield (kg/gross ha) Rice	1169	1382	18.22
Wheat	924	2191	139.72
Total	1166	1425	22.21
19 Foodgrain yield (kg/net ha)[d] Rice	1713	2118	23.70
Total	1708	2183	27.81

* 1969–70 only.
[a] Average of 1967–8, 1968–9 and 1969–70.
[b] Average of 1982–3, 1983–4 and 1984–5.
[c] Irrigation by shallow and deep tubewells, low lift pumps, and large-scale canals.
[d] Adjusted for multiple cropping (inflated by intensity of cropping).

Source: Based on data from sources mentioned in Table 2.1.

2.4.1   An Overview of Technological Change: 1967–8 to 1984–5

Here, by way of necessary background, selected indicators of changes in Bangladeshi agriculture since the advent of the new technology are presented. Changes between two three-year average end-points in time are compared. The first point is 1967–9 (average for the years 1967–8, 1968–9 and 1969–70) while the second is designated 1982–4 (average for the years 1982–3, 1983–4 and 1984–5).

The data pesented in Table 2.2 reveal the following changes between the two end-points:

1. Total cropped area (including multiple cropping) increased marginally while net cultivated are declined by 1.7 per cent, with an increase in cropping intensity from 146.5 per cent to 153.2 per cent between 1967–9 and 1982–4.
2. Gross area under rice cultivation increased by 4.72 per cent. The most significant change has taken place in wheat area, which rose spectacularly by nearly 450 per cent. The relative share of cereals in gross cropped area increased by 6.4 per cent (from 78.4 per cent in 1967–9 to 83.4 per cent in 1982–4), while that of non-cereals declined by 23 per cent (from 21.6 per cent to 16.6 per cent) in the same period.
3. Area under HYV rice rose significantly from 1.6 per cent of total rice area in 1967–9 to nearly 26 per cent in 1982–4. As of 1982–4, practically all the wheat area is under HYV cultivation.
4. Total area irrigated increased significantly (from 8 per cent of gross cropped area in 1967–9 to 14.71 per cent in 1982–4). The bulk of the irrigated area is confined to the production of rice and wheat. The share of modern irrigation rose from less than a third of the total to more than two-thirds between 1967–9 and 1982–4.
5. Use of chemical fertilizers (in nutrient terms) per gross cropped hectare rose by a factor of four.
6. Overall food production increased by 33 per cent, rice output increased moderately by 24 per cent and wheat by 1200 per cent. Foodgrain yields on the whole rose by about 22 per cent, while those of rice and wheat increased respectively by about 18 per cent and 140 per cent. Yields per net cropped hectare of foodgrains and rice rose respectively by nearly 28 per cent and 24 per cent.

2.4.2 Output and Productivity Implications of Technical Change

In this section, an analysis is undertaken of the growth in output and yield of Bangladeshi crops by major commodity groups in the period of the new technology. The growth rates are estimated employing Equation (2.3) which involves a single kink and index numbers of the relevant variables, with the average of triennium ending 1977–8 as the base. The index numbers for output and yield for the commodity groups, Cereals, Cash crops and Oilseeds are the same as before, while those for Pulses, Fruits, Vegetables and Spices are derived on the basis of data from sources mentioned in Tables 2.1 to 2.8. As before, a trend break is assumed at 1974–5 which marks the end of the first part of the period of the new technology.

Results of this exercise are set out in Table 2.3. When the growth rates for comparable crops, e.g. Cereals, are compared, the growth rates in the second period are slightly higher. The estimates in Table 2.3 are derived on the basis of a fewer number of observations. Because of the inclusion of two more crops in the commodity group Pulses it can be seen that the output growth rates in the last two sub-periods differ. Whereas in Table 2.1 output of Pulses is estimated to grow at a very low negative rate, it becomes strongly and statistically significantly negative in Table 2.3. For Oilseeds the overall pattern does not change but differences occur because of the differences in the total number of observations involved. The growth rate of non-cereals seems to be strongly negative in the 1967–74 period in Table 2.3 while the positive growth rate in the 1975–84 period becomes stronger and statistically superior with the inclusion of additional crops. Fruits, Vegetables and Spices all record higher growth rates in the last sub-period relative to the earlier one. Vegetables experience significantly positive output growth rate in the later period. Overall, however, growth rates in the output of 36 crops are marginally lower than those of the 21 crops considered earlier.

Growth in yields per hectare shows a broadly similar pattern to the one emerging from an analysis of Table 2.1. The net yields for all crops in Table 2.3 show relatively higher rates of increase than those in the 1975–84 period. However, lower rates of growth are seen to occur in the 1967–74 period when greater numbers of crops are considered. These suggest relatively higher decline or lower rates of productivity growth of the additional crops in the relevant period.

Thus even though one needs to be wary of the quality of estimated growth rates for some of the crops because of poor explanatory power and statistical significance of the coefficients, there is no evidence to suggest that non-cereal output has shown any strong tendency to increase. One implication of this behaviour of crop output growth is the change in overall crop output-mix. The share of cereals in total crop output has increased from 72 per cent in 1967–9 to 80 per cent in 1982–4, suggesting a decline in the relative share of non-cereals from about 28 per cent to 20 per cent during the same period. Within the commodity groups there has been a significant change in the output-mix. For instance, in Pulses, *khesari* has become an increasingly important crop in the group while the importance of lentils and *moong* is on the decline.

Table 2.3   Estimated growth rates of output and yield of major crops, using a one-kink model: Bangladesh, 1967–8 to 1984–5

Crop	Intercept	1967–8 to 1974–5 Growth rate (%)	t-value	1975–6 to 1984–5 Growth rate (%)	t-value	R^2	F-Ratio
(a) Output							
Cereals	4.43366	1.66	2.83[a]	3.08	4.11[a]	0.7679	24.819[a]
Rice	4.44554	1.49	2.58[b]	2.42	3.27[a]	0.7001	17.511[a]
Wheat	3.30296	15.22	6.04[a]	24.53	7.58[a]	0.9267	94.796[a]
Cash crops	4.75045	−1.64	1.66[c]	1.56	1.23[d]	0.1651	1.483[d]
Jute	4.91045	−3.17	2.06[b]	0.92	0.47[d]	0.2391	2.357[c]
Oilseeds	4.73539	−1.85	2.91[a]	0.97	1.19[d]	0.3646	4.303[b]
Pulses	4.80096	−2.55	3.93[a]	−1.28	1.54[c]	0.6831	16.169[a]
Vegetables	4.57511	0.32	0.56[d]	3.26	4.46[a]	0.6715	15.328[a]
Spices	4.80326	−2.47	4.80[a]	−0.05	0.08[d]	0.6782	15.809[a]
Fruits	4.88478	−3.55	8.62[a]	0.56	1.06[d]	0.8562	44.642[a]
Non-cereals	4.77020	−2.00	3.85[a]	1.20	1.80[b]	0.4971	7.414[a]
All crops	4.51737	0.74	1.37[d]	2.69	3.86[a]	0.6655	14.923[a]
(b) Yield (per gross cropped hectare)							
Cereals	4.48706	1.09	2.76[a]	2.02	3.97[a]	0.7565	23.296[a]
Rice	4.49136	1.05	2.69[a]	1.93	3.83[a]	0.7440	21.798[a]
Wheat	3.88521	7.04	5.14[a]	5.77	3.28[a]	0.8299	36.585[a]
Cash crops	4.25701	1.54	2.79[a]	1.89	2.67[a]	0.6691	15.168[a]
Jute	4.50461	0.94	1.53[d]	1.89	2.40[b]	0.5163	8.004[a]
Oilseeds	4.70144	−1.09	3.35[a]	1.32	3.15[a]	0.4836	7.023[a]
Pulses	4.73451	−1.72	4.40[a]	0.05	0.10[d]	0.6293	12.730[a]
Vegetables	4.64538	−0.49	2.29[b]	1.58	5.70[a]	0.6877	16.516[a]
Spices	4.75124	−1.68	5.57[a]	−0.07	0.18[d]	0.7424	21.616[a]
Fruits	4.93710	−4.06	14.95[a]	−0.71	2.04[b]	0.9588	174.499[a]
Non-cereals	4.52895	−0.26	0.85[d]	1.41	3.59[a]	0.4855	7.077[a]
All crops	4.52228	0.57	1.69[c]	1.82	4.22[a]	0.7151	18.822[a]
(c) Yield (per net cropped hectare)							
Rice	4.47656	1.23	21.7[b]	2.98	4.09[a]	0.7325	20.532[a]
Cereals	4.47216	1.27	2.22[b]	3.07	4.18[a]	0.7413	21.486[a]
All crops	4.51989	0.75	1.44[c]	2.86	4.30[a]	0.7077	18.155[a]

[a] Significant at 1 per cent level;
[b] Significant at 5 per cent level;
[c] Significant at 10 per cent level;
[d] Not significant.

Source: Based on data from sources mentioned in Table 2.1.

2.5   DECOMPOSITION OF CROP OUTPUT GROWTH: 1967–9 TO 1982–4

Given technological change in Bangladeshi agriculture and the resulting increase in agricultural production, it is worth while trying to identify apparent sources of growth. To what extent have technological changes, in fact, raised the yields of crops? To what extent have they extended the area under production of particular crops? To what extent have yield effects and area effects contributed to the growth of agricultural output? Has the Green Revolution, after all, been very effective in raising yields? In this chapter decomposition methods are considered as a possible approach to answering them.

Several methods of decomposing agricultural output growth have been suggested. They can be broadly classified as production function and non-production function approaches. The former can sometimes be employed to estimate factor shares and identify sources of productivity growth (Johansen, 1972), though the non-production function approach of Minhas and Vaidyanathan (1965), which is reminiscent of Kendrick's (1956; 1961) and Denison's (1962), is more widely used (see, for example, M. Hossain, 1979; 1984; Narain, 1977; Pray, 1979; Sagar, 1977; 1980; Venkataramanan and Prahladachar, 1980). It divides growth in crop production into components due to alterations in (1) area, (2) yield, (3) cropping pattern, and (4) interactions of changes in cropping pattern and yield, and identifies the relative importance of each of these components. Wennergren *et al.* (1984) adapt the M-V (Minhas-Vaidyanathan) method in order to identify the contributions of area, yield, cropping pattern and interactions to growth in output of individual crops in Bangladesh. Venegas and Ruttan (1964) also employ a method which partitions output growth of a single crop into changes attributable to area and to yield.

This section critically reviews these three decomposition methods. It identifies a number of shortcomings and suggests that inappropriate economic inferences have been drawn from them. After the basic mathematics of the methods are sketched, they are used to decompose the growth in crop output in Bangladesh. Shortcomings are then analysed and their significant limitations are illustrated by data from Bangladesh which highlight the arbitrary nature of assumptions underlying the methods and the sensitivity of decomposition results to aggregation. It is contended that decomposition methods fail to identify accurately the sources of agricultural output growth and can give a misleading view of the impact of the new technology associated with the Green Revolution.

2.5.1 Decomposition Methods: Basic Mathematics

Venegas-Ruttan (V-R) method: This method examines the relationship between growth in aggregate output of an individual crop, area planted and yield per unit area. The analytical framework depends on the identity which expresses output (P) as the product of area planted (A) and yield (Y). Thus $P = AY$. Using subscripts 0 and t to represent a base and a terminal period respectively, the relationship between the two production levels can be expressed as

$$P_t/P_0 = (A_t Y_t)/(A_0 Y_0) = (A_t/A_0) \cdot (Y_t/Y_0) \qquad (2.7)$$

The above relationship is used as a means of deriving implications for changes in total output in terms of changes in land productivity and area under cultivation. Taking logarithms of both sides of (2.7) and making rearrangements, one gets

$$[ln(A_t/A_0)/ln(P_t/P_0)] \times 100 + [ln(Y_t/Y_0)/ln(P_t/P_0)]$$
$$\times 100 = 100\% \qquad (2.8)$$

where the first term in (2.8) represents the percentage change in total production due to change in area, and the second term represents the yield effect. Thus all variation in output is decomposed into sums attributed to area and yield. Venegas and Ruttan (1964) applied this method to determine the relative contribution of area and yield to increased rice production in the Philippines.[3]

Robert Niehaus of the Economic Research Service, United States Department of Agriculture (USDA), has used essentially the same formula to partition the influences of area and yield on total crop production:

$$[\log(1 + a)/\log(1 + p)] \times 100 + [\log(1 + y)/\log(1 + p)]$$
$$\times 100 = 100\%$$

where a, y and p are percentage changes in area, yield and production respectively (see Dalrymple, 1977, p. 187).

The assumptions underlying this decomposition method are not explicitly stated. However, yield per hectare as an index of average land productivity conceals an important aspect of agricultural output growth since it does not take account of differences in land quality.

Some implications of this method can be derived as follows. For small changes:

$$\Delta \log A / \Delta \log P \simeq \delta \log A / \delta \log P = (\delta A / \delta P) \cdot (P/A) = 1/e_a \quad (2.8a)$$

and

$$\Delta \log Y / \Delta \log P \simeq \delta \log Y / \delta \log P = (\delta Y / \delta P) \cdot (P/Y) = 1/e_y \quad (2.8b)$$

where e_a and e_y can be called area and yield elasticities respectively. Taking the identity $P=AY$, in the limiting case (e.g. with small changes), the reciprocals of the two elasticities approximately sum to unity. This appears to be analogous to the special case of the Cobb-Douglas production function when the sum of the exponents equals unity, thereby displaying constant returns to scale.

Alternatively, however, apart from area and yield effects there can be a third effect: a mixed effect which is a combination of the two. Given that $P=AY$, ΔP can be decomposed as follows:

$$\Delta P = (\Delta A)Y \quad + \quad (\Delta Y)A \quad + \quad (\Delta A)(\Delta Y) \quad (2.8c)$$

Pure area effect   Pure yield effect   Mixed effect

If this method of decomposition is used, it will usually give a different breakdown of contributions from area and yield. For example, if area and yield have increased, this would imply a positive mixed effect and the pure yield effect must be smaller than the yield effect suggested by the V-R method.

Minhas-Vaidyanathan (M-V) method: Minhas and Vaidyanathan (1965) employ a decomposition technique to analyse aggregate crop output growth by component elements. They take a group of crops and decompose the growth of total crop output into the contribution of (1) changes in area, (2) changes in yield, (3) changes in cropping pattern, and (4) interaction between changes in cropping pattern and yield. Let P = total value of crop output, Y_i = yield of the ith crop, and C_i = proportion of the gross cropped area A under the ith crop. Define:

$$P_0 = A_0 \Sigma W_i C_{i0} Y_{i0} \quad (2.9)$$

$$P_t = A_t \ \Sigma W_i \ C_{it} \ Y_{it}. \tag{2.10}$$

where W_i refers to constant price weights assigned to the ith crop, and 0 and t respectively refer to base and terminal periods. Then,

$$(P_t - P_0) = (A_t - A_0) \ \Sigma W_i \ C_{i0} \ Y_{i0} + A_t \ \Sigma W_i \ (Y_{it} - Y_{i0}) \ C_{i0}$$
$$+ \ A_t \ \Sigma W_i \ (C_{it} - C_{i0}) \ Y_{i0}$$
$$+ \ A_t \ \Sigma W_i \ (C_{it} - C_{i0}) \ (Y_{it} - Y_{i0}). \tag{2.11}$$

The *first* term on the right-hand side of equation (2.11) represents the area effect, and shows the magnitude of the output increase that would have taken place *if there were no change in yield and cropping pattern*. The *second* term is the effect of change in yield for a constant cropping pattern. The *third* term represents the effect of change in cropping pattern in the absence of any change in yield. The *last* term is the output change attributable to interaction between changes in yields and cropping patterns. The interaction term 'is essentially in the nature of a balancing entry' (p.236). With the aid of these equations, the relative importance of each can be ascertained. Minhas and Vaidyanathan applied their method to analyse growth of crop output in India. In recent years, M. Hossain (1979; 1980; 1984) and Pray (1979) have applied it to 'identify' the sources of growth of Bangladeshi crop output. A crucial underlying assumption is that 'every new gross crop acre is as good as an acre already under cultivation' (Minhas and Vaidyanathan, 1965, p.234). This implies constant returns.

Wennergren, Antholt and Whitaker (W-A-W) method: This method (Wennergren *et al.*, 1984) is essentially an extension of the M-V method. While the latter concentrates on the changes in aggregate crop output, the former estimates the same components of growth in production of individual crops. For a single crop, let P = output, C = proportion of the gross cropped area under the crop, A = crop area under particular crop, and Y = yield. The components 'explaining' the changes in the total production of individual crops between year 0 and year t are algebraically as follows:

Change in total output $(P_t - P_0) =$
Area effect (V) $Y_0[A_t(1 + C_0 - C_t) - A_0]$ $+$

Yield effect (S) $[A_t(1 + C_0 - C_t) (Y_t - Y_0)] +$
Cropping pattern (R) $[A_t Y_0(C_t - C_0)]$ $+$
Interactions (U) $[A_t (Y_t - Y_0)(C_t - C_0)]$ (2.12)

This method decomposes individual crop output growth into more components than that of Venegas and Ruttan. In addition to area and yield it includes cropping pattern and interaction components. However, the crucial assumptions underlying the W-A-W method are those of Minhas and Vaidyanathan and imply constant returns. Some further implications of the underlying assumptions will be examined in Section 2.6.

2.5.2 Some Empirical Results

The decomposition schemes mentioned in the previous section take the change in output (aggregate of crops or an individual crop) as given, and measure the relative importance of the component elements. Taking into account the limitations of growth rates between the two end-points mentioned in Section 2.2, changes in crop output growth of Bangladesh between the base period 1967–9 and a terminal period 1982–4 are decomposed into the elements suggested by the various approaches.

Table 2.4 presents results for the V-R method. It can be seen that both area and yield declines 'explain' the fall in output for local varieties of *aus, aman* and *boro* rice and wheat, area reduction being dominant. For local *aman*, the output fall is almost entirely accounted for by area decline which is only slightly offset by an insignificant increase in yield. The output growth of all HYV rice (which is huge in the case of *aman*) results primarily from an area increase which is partially offset by a decline in yield. Increase in area with a minor yield effect 'explain' HYV wheat output changes, which are almost as dramatic as those for *aman* rice. For jute, a positive yield effect partially offsets a strong negative area effect. For tobacco, a stronger positive area effect supplemented by a positive yield effect contributes to output increase. For sugar cane, output decline is accounted for by decline in yield which is only partially offset by positive area effect. For oilseeds, a positive yield effect partially offsets a negative area effect, while both effects are negative in the case of pulses. Output of fruits declines primarily as a result of negative yield effect. Vegetables show positive output change following area increase. Both area and yield declines account for negative output growth of spices.

Table 2.4 provides decomposition results for individual crop output growth using the W-A-W method, thus involving more components than the V-R method. In addition to area and yield it includes cropping pattern and interaction components. The pure yield effect is negative for all crops apart from both varieties of wheat, jute, tobacco and oilseeds. Area and cropping pattern effects 'explain' the output decline of relevant crops. For HYVs of rice, positive area and cropping pattern effects 'explain' output growth, although negative yield effects tend partially to offset them.

Results of an exercise based on the M-V method to 'identify' sources of aggregate crop output growth are presented in Table 2.5. The analysis involves the 36 crops mentioned earlier in this chapter. These results are derived using the 1972–3 and 1981–3 harvest prices of the relevant agricultural commodities. As the separate prices for local and HYVs of different cereals are not available, the same price is used for the derivation of values of output of local and HYV cereals. Because of different weights used during 1981–3, slight differences in the local and HYV cereal prices occur. Where aggregation is involved, e.g. for rice, wheat, pulses, etc., the average prices are derived as weighted averages, with the component production figures used as weights. It can be seen that only area increase and cropping pattern have had a positive impact on total crop output growth. The pure yield effect and the interaction of cropping pattern and yields are negative and, therefore, partially offset the positive contributions of area and cropping patterns. The most significant contribution to output increase is made by changes in cropping pattern. The results based on two sets of prices in Table 2.5 show very little difference, but those based on the 1981–3 prices tend to produce an underestimation of the relative importance of the component elements in either direction.

2.6 SOME SPECIFIC THEORETICAL SHORTCOMINGS OF THE M-V TYPE DECOMPOSITION SCHEME

The development of decomposition methods for analysing sources of agricultural productivity growth is an understandable response to shortages of and deficiencies in data in LDCs which limit the scope

Table 2.4 Growth of output of selected crops, Bangladesh, 1967–9 to 1982–4: Analysis by component elements, using the V-R and W-A-W methods of decomposition

Crop	V-R Method % Change in output	V-R Method % Change due to Area	V-R Method % Change due to Yield	V-R Method LI term*	V-R Method Total	W-A-W Method Area	W-A-W Method Sources of growth Yield	W-A-W Method Sources of growth Cropping pattern	W-A-W Method Interactions
Aus local	-29.30	66.80	33.20	0.00	100	53.47	31.30	17.09	-1.86
Aus HYV	3272.53	115.36	-15.36	0.00	100	167.38	-71.17	6.50	-2.71
Aman local	-19.55	93.79	6.21	0.00	100	51.03	6.18	43.37	-0.58
Aman HYV	13116.60	107.35	-7.36	0.01	100	131.57	-39.93	11.98	-3.61
Boro local	-47.89	91.70	8.30	0.00	100	91.21	6.19	2.74	-0.14
Boro HYV	460.95	117.74	-17.74	0.00	100	131.05	-40.25	12.50	-3.29
Wheat local	-78.32	104.65	-4.66	0.01	100	101.74	-1.91	0.17	0.01
Wheat HYV	12589.60	92.25	7.75	0.00	100	65.50	30.20	2.95	1.34
Jute	-25.13	163.45	-63.45	0.00	100	142.40	-51.51	7.58	1.53
Sugar cane	-5.55	-12.91	112.91	0.00	100	-14.11	113.38	0.78	-0.05
Tobacco	22.42	65.74	34.26	0.00	100	63.26	36.55	0.18	0.01
Pulses	-24.61	72.99	27.00	0.01	100	73.78	24.43	1.934	-0.14
Oilseeds	-11.51	124.49	-24.49	0.00	100	119.08	-22.79	3.60	0.11
Vegetables	15.95	100.10	-0.11	0.01	100	98.96	-0.11	1.15	0.00
Fruits	-20.90	-46.46	146.46	0.00	100	-54.73	154.97	-0.34	0.10
Spices	-22.39	34.24	65.76	0.00	100	36.55	62.96	0.58	-0.09

* Logarithmic interpolation term.

Source: Based on data from sources mentioned in Table 2.1.

Table 2.5 Contribution of various components to growth of aggregate crop output growth: Bangladesh, 1967–9 to 1982–4, using the M-V method of decomposition

Sources of growth	Percentage change in crop output		Relative contribution (%)	
	1981–3 price	1972–3 price	1981–3 price	1972–3 price
Area	4.09	4.09	21.63	20.48
Yield	–5.26	–4.46	–27.82	–22.33
Cropping pattern	33.82	35.26	178.85	176.56
Interactions	–13.74	–14.92	–72.66	–74.71
Total	18.91	19.97	100.00	100.00

Note: Crops include six varieties of rice, two varieties of wheat, jute, sugar-cane, tobacco, six varieties of pulses, five varieties of oilseeds, four crops of fruits, five crops of vegetables and four varieties of spices mentioned in the text. Same price for local and HYVs of cereals has been used. 1981–3 prices are weighted averages for the years 1981–2, 1982–3 and 1983–4 with respective quantities used as weights, while 1972–3 prices are those of a single year. To derive the output value of rice, a rice–paddy conversion ratio of 0.67 has been used. It normally ranges between 0.67 and 0.70 – M. Hossain (1980, p.54) uses 0.70.

Source: Based on data from sources mentioned in Table 2.1.

for estimating production functions (Minhas and Vaidyanathan, 1965). Nevertheless, there are considerable dangers involved in using these methods. Their use can easily lead to unwarranted conclusions and assessments and become a source of faulty policy advice.

While Minhas and Vaidyanathan point out that their decomposition scheme is arbitrary, even they read too much significance into results obtained by applying it. They also mention that in their scheme, 'component elements are so chosen that their contributions to output growth are determined by more or less independent sets of factors. Each of these sets of factors can be separately analysed and these analyses should provide the building blocks for constructing output predictions' (p.235). However, the degree of interdependence of components will be significant if price effects are important and changes in relative prices occur. A rise in the price of a particular crop can lead to a rise in area planted to it relative to other crops. In the absence of any other change, and given diminishing marginal productivity, the average physical yield of the crop with relative rise in price will fall, while that for the crops for which it substitutes is

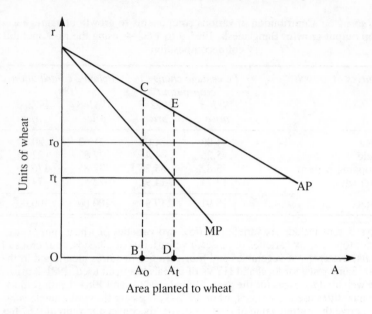

Figure 2.2   Price effects on yield limit the validity of the Minhas-Vaidyanathan method

liable to rise. This is illustrated in Figure 2.2 by a hypothetical case. A rise in the real value of land for, say, growing wheat relative to other crops might, other things unchanged, result in a reduction in the real (opportunity) cost of using land for wheat. This, for instance, will occur if the price received for produce other than wheat falls and other things remain unchanged. In Figure 2.2, for instance, the relative cost of using land for growing wheat might fall from r_1 to r_t because the price received for produce other than wheat falls. Given that *AP* represents the average product curve for wheat production and *MP* represents the marginal product, price variation results in the area being used for wheat expanding from A_0 to A_t and the average product or yield of wheat falling from *BC* to *DE*. It should be noted that this is not indicative of technological change (a point taken up later) and that area and yield are not independent in this case. Furthermore, the cropping pattern is not independent of the other components either. Thus, when price variations are significant, the Minhas-Vaidyanathan claim for independence of components is not justified.

Even greater claims have been made for the significance of the method by later authors. They seem, however, to have been lured into spurious quantification. While Pray (1979) is relatively careful in applying the M-V method to Bangladesh, later writers such as Wennergren *et al.* (1984, p.86) seem to be less moderate in their claim and assessments. The latter claim that differences in yields obtained using the M-V method 'indicate the degree to which adoption of new technologies has influenced production' (p.86). They use the modified method to assign sources to changes in production of major crops in Bangladesh between 1973–4 and 1981–2. A large fraction of the increased production is assigned to the area factor rather than to yield increases.

In order to appreciate some pitfalls of the method, one can concentrate on the simple one-product case. This involves decomposition only into yield and area effects. It is assumed here that $C_t = C_0$. The limitations that show up will also apply to the M-V aggregative approach. Suppose that yield of crop in the base year is Y_0 and area A_0 and that these rise by the year t to Y_t and A_t respectively. In these circumstances, according to the Wennergren *et al.* modification of the M-V formula, the change in total output of the crop can be decomposed as follows:

$P_t - P_0$ (change in total output)

$= Y_0(A_t - A_0)$ (change in area effect)

$+ A_t(Y_t - Y_0)$ (change in yield effect)   (2.13)

This decomposition is illustrated in Figure 2.3. Output increases from the area shown by the rectangle $OA_0\,GY_0$ to that by rectangle $OA_t\,JY_t$. The output increase is decomposed into that due to area (the area of $A_0\,A_tHG$) and that due to yield variation (the area of $Y_0\,HJY_t$). The values assigned to yield come out as a remainder because it is assumed that all additional area used in growing the crop has the same yield on average as that used to grow the crop in the base year if the techniques of the base year are used (Minhas and Vaidyanathan, 1965, p. 234). This indicates that constant marginal productivity (or returns) is assumed. This assumption does not seem to be characteristic of agriculture. One wonders why the additional land was not used to grow the crop if it was in fact as productive as land already under use.

The constant marginal productivity assumption is a crucial one.

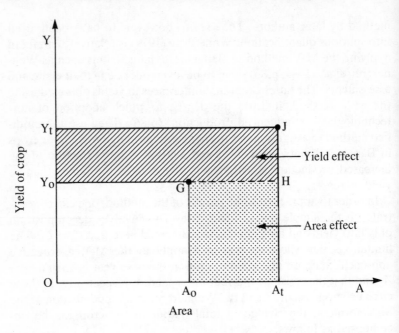

Figure 2.3   Illustration of decomposition into yield effect and area effect of an increase in output of a crop using the Wennergren *et al.* method

What if the yield on the additional area were lower than Y_0? Suppose for instance it happened to be $0.5Y_0$. Then using the logic of the M-V analysis the area effect accounts for only $0.5Y_0(A_t - A_0)$ of the increase in production, and the yield effect accounts for the remainder. It is even conceivable that, using base technology, the yield on the additional area would be zero. In that case all of the increased production of the crop should be assigned to the yield effect. In the absence of technological progress the additional area may not be available (it may only come as a result of an increase in possibilities for multiple cropping stemming from technological progress). It could then be argued that the whole output increase should be seen as a yield increase. If one is applying this method one should be careful to make appropriate assumptions about the productivity of additional area and not blindly follow the Minhas-Vaidyanathan assumptions. Earlier studies (e.g. M. Hossain, 1979; 1984; Pray, 1979; Wennergren *et al.*, 1984) have not done this. Assuming that diminishing marginal productivity (and average productivity) is the rule for most products as area is expanded using constant technology,

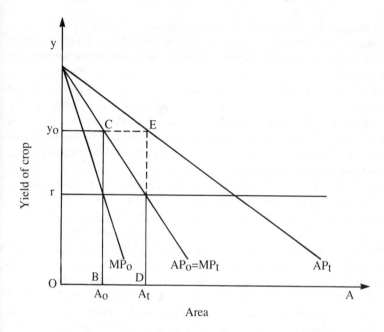

Figure 2.4   Illustration of a case in which average yield of a crop remains unchanged even though technological progress increases productivity

the M-V method overstates the significance of area effects and understates the contribution of yield effects.

Absurd conclusions can be drawn from the analysis if variation in yields is used as an indicator of the impact of technological progress on production. Firstly, there may be no technological progress and yet yields may rise. Secondly, technological progress may contribute unequivocally to increased production and yet yield per hectare may remain unchanged or even decline. Each of these propositions is considered in turn.

In Figure 2.2, suppose that the real cost of production of the crop rises from r_t to r_0, area falls from A_t to A_0 and yield rises from DE to BC. The M-V method would assign a proportion of change to area reduction and a proportion to yield increases. In this case, however, the increase in yield is entirely due to the presence of a price effect given diminishing marginal productivity, and not to technological change. The rise in yield is not associated with any new technology or change in production function. This illustrates the first proposition.

Figure 2.4 illustrates a case in which average yields remain unchanged as technological progress occurs. In this case there is, however, an upward movement of the production function which makes all land used for the crop in the base period more productive. In Figure 2.4 the downward slope of the average product curve in period t is half that in the base period. AP_t represents the yield or average product curve for the crop in period t and AP_0 that in the base period. The corresponding marginal product curves are MP_0 and $MP_t = AP_0$. Assuming that the real cost per hectare of land remains unchanged at r, area in the base period is A_0 and subsequently A_t, assuming profit maximization. Yield is BC in period 0 and DE in period t but $BC = DE$: yield is constant. Therefore, the M-V method would assign the whole production increase $(A_t - A_0)Y_0$ to the area effect. Yet as can be seen from the shift in the marginal productivity curve from MP_0 to MP_t, productivity increased on every hectare of land devoted to the crop compared to actual (or hypothetical) productivity in the base period. The production function has shifted upwards, except at zero output, as a result of technological progress.

The matter can also be considered by direct reference to production functions. In Figure 2.5, suppose the production function for a crop is as indicated by the curve marked $f_0(A)$ and that by period t it has shifted to $f_t(A)$ as a result of technological progress. If area expands from A_0 to A_t, in period t the average yield of the crop remains unchanged as shown by the slope of lines OL and OM. The M-V method assigns the whole impact of production change to an area affect. Even worse! If area expands say, to A_t', yield falls as shown by the slope of line ON in relation to that of OL. Consequently the M-V method decomposes the total production change into a negative yield effect and an even larger positive area effect. This is patently absurd.

Apart from the difficulties mentioned above, variations in relative availability of resources (e.g. fertilizer prices) may compound problems of identification. Given climatic and other environmental variabilities, the question arises: How representative are the base year and the terminal year? To what extent have yields in particular been influenced by 'random' environmental factors? (cf. Wennergren et al. 1984, p. 122). It can be misleading to apply the method mechanistically.

The yield of a crop can alter even though its production function remains unaltered, as was pointed out in relation to Figure 2.2. In this case, it was pointed out that a rise in the relative price of, say, wheat would increase wheat acreage and reduce the average product

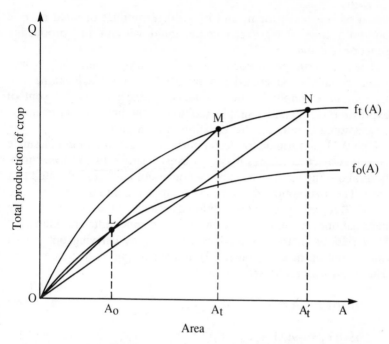

Figure 2.5 Further illustration of a case in which average yield of a crop remains unchanged as production function moves upwards, together with a case in which the average yield actually falls

(yield) of wheat. At the same time, land is likely to be withdrawn from other crops and their yield (average product) rises, given that average product declines when greater acreage is devoted to a crop. Such changes can come about as a result of altered patterns of demand, and occur in the absence of any variation in production. Furthermore, even if area allocated to a crop remains unaltered, the composition of land used for its production may have altered and this can affect conclusions from the M-V analysis. For example, one type of land may move out of a crop because of a rise in the price of an alternative crop and *another type* of land may be switched into production of the crop because of the fall in the price of an alternative. It is theoretically possible for these movements to cancel out and for the acreage of the crop to remain unaltered. The product of the crop at the margin may be the same, but below the margin the marginal product may have altered because of the changed composition of land used for it, and its average product or yield may be altered. These changes can come about without the production function shifting.

The limitations illustrated for a particular crop generalize to aggregate production as considered by Minhas and Vaidyanathan. If $C_t = C_0$, for example, interactions and cropping patterns 'drop out' of expression (2.11) and a simple addition of the first two terms in the expression accounts for the variation in total output.

The W-A-W formula for decomposing changes in output of a single crop is subject to various biases. It is especially sensitive to assumptions about the yield to be expected on additional area allocated to a crop. The assumption about Y_0 (in this case the initial value and also the yield to be expected on additional area allocated to a crop) is the material one. It can be seen how various components are influenced by variations in the assumed values of Y_0 by carrying out partial differentiation of equation (2.12) and supposing Y_t to be constant. The following results are obtained:

$$\delta V / \delta Y_0 = A_t(1 + C_0 - C_t) - A_0, \tag{2.14}$$

$$\delta S / \delta Y_0 = -A_t(1 + C_0 - C_t), \tag{2.15}$$

$$\delta R / \delta Y_0 = A_t(C_t - C_0), \tag{2.16}$$

$$\delta U / \delta Y_0 = -A_t(C_t - C_0). \tag{2.17}$$

Usually the impact of assuming a greater initial yield is to reduce the amount of change attributed to the yield effect: that is, $\delta T / \delta Y_0 < 0$. In the particular case where $C_t = C_0$, (2.14) is positive if $A_t > A_0$, (2.15) is negative and the other expressions reduce to zero.

Yield may be biased upward because it is assumed that yield on existing area should be equal to that on additional area, whereas even under ideal conditions it may be less. The choice of prices can also introduce a bias. Minhas and Vaidyanathan (1965) assume prices of the base period in computing their results. In this respect their assumption is similar to that of Hicks (1940; 1948) concerning the valuation of national income, and differs from that of Kuznets (1948).[4] If later prices are used instead of base period ones and these are higher than those in the base period, this increases the assumed value of the initial yield and thus reduces the yield effect.

2.7   FURTHER OBSERVATIONS ON DECOMPOSITION METHODS USING BANGLADESHI DATA

This section uses Bangladeshi data to indicate how the different sources of growth of agricultural output can be influenced by (1) the level of aggregation of the data and (2) the set of prices used to calculate aggregate output. First, the sensitivity of the V-R and W-A-W methods and then the sensitivity of the M-V method to degree of aggregation of crop production is examined. Later the sensitivity of the M-V method to the price set used to determine the total value of the crop is considered.

Consider aggregate rice output (six varieties taken together) as well as rice production for *aus, aman* and *boro* individually and wheat output. Thus *aus* in Table 2.6 implies total output of local and HYV *aus* rice. Results after such aggregation are presented in Table 2.6. When compared with the information set out in Table 2.4, Table 2.6 shows that the relative importance of the assigned sources of growth is sensitive to levels of aggregation employed. Thus when rice as a whole or *aus* and *aman* rice are considered (Table 2.6, part a) it can be seen that the primary source of growth is the increase in yields, which does not appear to be the case when each of the individual varieties is considered (Table 2.4) because yield effects are negative and area effects are strongly positive for most varieties (in fact all HYVs except wheat) of these crops. For *boro* rice both effects are positive with area being dominant, but when HYV *boro* is considered, the yield effect is negative. For wheat, the impact of yield appears to be quite significant. A significant change in the relative importance can also be noticed when Table 2.4 and Table 2.6 (part b) are compared. The importance of the cropping pattern as an assigned source of growth relative to that of yields is much reduced on aggregation. The sensitivity of assigned sources of growth to aggregation is likely to provide conflicting evidence about the actual relative sources of growth of crop output. However, one might note from parts (a) and (b) of Table 2.6 the closeness of the area and yield effects as measured by the two methods.

In comparing the growth of aggregated crop production in Table 2.6 with that of selected single crops in Table 2.4, it can be seen that the yield effect is a relatively more significant source of growth of output in the aggregated case. This suggests that the main significance of the Green Revolution lies not so much in its raising the yield

Table 2.6   Percentage influence of various component elements of individual crop output growth: Bangladesh, 1967–9 to 1982–4

(a) V-R method Crop	Percentage change in production	Percentage of change due to			Total
		Area	Yield	LI term*	
Rice	23.82	21.28	78.72	0.00	100
Aus	1.42	–456.81	556.81	0.00	100
Aman	9.88	–4.80	104.80	0.00	100
Boro	129.74	77.20	22.81	–0.01	100
Wheat	1196.44	66.29	33.71	0.00	100

(b) W-A-W method Crop	Percentage change in production	Relative contribution of sources of change				Total
		Area	Yield	Cropping pattern	Interactions	
Rice	23.82	17.61	80.12	1.92	0.35	100
Aus	1.42	–265.08	553.52	–174.20	–14.24	100
Aman	9.88	16.55	106.76	–21.11	–2.19	100
Boro	129.74	61.83	29.01	7.58	1.58	100
Wheat	1196.44	35.68	60.41	1.65	2.27	100

* Logarithmic interpolation term.

Note: Respective components may not always add up to the total because of rounding.

Source: Based on data from sources mentioned in Table 2.1

from single cropping but in its increasing the incidence of multiple cropping and thereby raising annual yields.

Earlier application of the M-V method considered outputs, yields and cropping patterns of each individual crop and variety separately, using their respective prices to get an overall picture. The assigned sources of growth presented in Table 2.5 were derived on this basis. The estimates of the sources of growth presented in Table 2.7 are derived after aggregating crop production data for local and HYVs of individual cereals. It involves aggregated output of *aus, aman* and *boro* rice and wheat and that of all other non-cereal crops mentioned in Table 2.4. Results employing further aggregation – that is, rice considered as a single crop along with all other crops mentioned earlier – are also presented using the 1972–3 and 1981–3 prices.

The impact of changes in price and aggregation on the relative

Table 2.7   Sensitivity to changes in price and level of aggregation of various sources of crop output growth: Bangladesh, 1967–9 to 1982–4, using the M-V method of decomposition

| Sources of growth | Relative importance (%) | | | |
| | Aggregation I | | Aggregation II | |
	1981–3 Price	1972–3 Price	1981–3 Price	1972–3 Price
Area	21.75	20.48	22.27	21.89
Yield	41.00	44.87	70.71	76.83
Cropping pattern	19.22	19.68	–4.04	–6.47
Interactions	18.03	14.97	11.06	7.75
Total	100.00	100.00	100.00	100.00

Notes: Under Aggregation I crops include three varieties of rice (*aus*, *aman* and *boro*), wheat and all the non-cereals mentioned earlier. *Aus* refers to the total of local and HYVs. The same applies to *aman*, *boro* and wheat crops. Under Aggregation II crops include aggregate of all the rice crops, wheat and all the non-cereals mentioned earlier.

Source: Based on data from sources mentioned in Table 2.1.

importance of various component factors in aggregate crop output growth can be seen from a comparison of Table 2.5 and Table 2.7. Table 2.5 suggests area and cropping pattern are major sources of growth, with yield and interactions having a negative impact. The results do not appear to be significantly influenced by changes in price weights. But when aggregation is employed, all the sources show positive contributions. The yield effect is the most dominant source of growth, followed by the area effect and interactions. The change in apparent relative sources of output growth when the set of prices used to value agricultural output is altered appears to be larger when greater aggregation is employed. Thus with the highest degree of aggregation, yield effect assumes overwhelming importance. At least this is so on the basis of the Bangladeshi data considered here. The higher percentage contribution of yield effect in terms of 1972–3 prices is probably due to (1) relatively faster increase in the prices of non-cereals, and (2) declining yields of non-cereals, particularly of pulses, fruits and spices.

So far decomposition of crop output growth has been considered in terms of gross cropped area, including multiple cropping. Even though, with the employment of greater aggregation, the yield effect assumes greater significance and is identified as the primary source of

Table 2.8   Sensitivity to adjustment for multiple cropping of various components of aggregate rice production and crop output growth, using V-R, W-A-W and M-V methods of decomposition

(a) *Rice production*

(i) V-R method

| Area | Percentage of change due to | | Total |
	Yield	LI term*	
0.36	99.64	0.00	100

(ii) W-A-W method

Percentage contribution of sources of growth

Area	Yield	Cropping pattern	Interactions	Total
1.62	99.99	−1.30	−0.31	100

(b) *Aggregate crop output growth*

(iii) M-V method

| Price weights | Relative Importance of components of growth | | | | Total |
	Area	Yield	Cropping pattern	Interactions	
1981–3	2.69	89.70	−2.50	10.11	100
1972–3	2.64	96.00	−5.60	6.95	100

* Logarithmic interpolation term.

Source: Based on data from sources mentioned in Table 2.1.

growth, it might still be under-represented with the area effect being over-represented in relative importance (Alauddin and Tisdell, 1986c). One way to overcome this is to use the *net cropped area* under cultivation of each crop. However, data on *net cropped* area are not readily available. The alternative is to use the intensity of cropping index for the two end-points as a deflator to derive an approximate estimate of net cropped area. However, apart from rice, none of the crops considered in the present analysis is grown twice on the same land during one year. Therefore the gross cropped area under rice (as a whole), deflated by the index of cropping intensity (1.465 during 1967–9 and 1.532 during 1982–4, Table 2.2), gives a reasonable approximation of the actual area under cultivation. The remainder of this section is devoted to an examination of the implications of such an adjustment for the assigned sources of growth. The results of this exercise are set out in Table 2.8.

First, V-R and W-A-W methods are applied to see the changes in the relative importance of various component elements contributing to overall rice output growth. It can be clearly seen that the yield effect almost entirely accounts for rice output growth. Other effects pale into insignificance. A similar picture emerges when the M-V method is applied to analyse aggregate crop output growth, using the adjusted rice area data. Furthermore, the difference in relative importance of yield as a factor of growth becomes more apparent as price weights are changed. While the yield effect accounts for 96 per cent of crop output growth in terms of 1972–3 prices, it explains about 90 per cent when 1981–3 prices are used.

2.8 SUMMARY AND CONCLUSIONS

A number of important changes have taken place in Bangladeshi agriculture since the introduction of the new agricultural technology. These include, among other things, the increase in the incidence of multiple cropping and decline in net cultivated area, changes in cropping pattern and output-mix. Overall crop output growth does not appear to be impressive, even though food production, while still not sufficient to feed the population (discussed in greater detail in Chapter 12), has increased significantly, primarily through a moderate increase in the output of rice and a spectacular boost in the production of wheat. The period under consideration also witnessed a trend away from non-cereal to cereal production. The output of non-cereal crops taken as a whole has not shown any tendency to increase. If anything, it has probably declined, and this decline has been the outcome of decline in yield as well as decline in hectarage.

Given these changes in the crop sector of Bangladeshi agriculture, significant production increases have taken place following the Green Revolution. However, as demonstrated, methods for decomposing agricultural productivity growth into its various 'sources' such as that developed by Minhas and Vaidyanathan (1965) and applied to Bangladesh by such writers as Wennergren *et al.* (1984) are likely to give a misleading picture of productivity effects of technological change. When the area planted to a particular crop of variety of a crop increases, and those decomposition methods are applied, there is a tendency to under-represent the yield effect if some fall in the average yield per hectare occurs, and to over-represent the area effect. Since the Green Revolution has increased the area under several varieties of crops (for example, dry season crops in Bangladesh), the yield effect often appears to be negative or underestimated by these methods. This can occur even when the productivity of the crop on *every* unit of land is raised. To say that yield decreases in such a case because *average* yield per hectare falls is not only peculiar but grossly misleading.

These decomposition methods mask an important effect of the Green Revolution on production. One of its most important impacts has been to increase the intensity of cultivation or cropping by increasing the incidence of multiple cropping, that is, the number of crops grown on cultivated land in each year. Decomposition methods take account of this as an increase in area devoted to particular crops.[5]

In Bangladesh, the area actually cultivated annually fell from around 8.8 million hectares in the late 1960s to around 8.6 million hectares in the early 1980s (BBS, 1978; 1985b; 1986d). The incidence of multiple cropping increased with the introduction of the new technology. For example, the area triple cropped (annually) expanded from around 480 to 649 thousand hectares between 1965–9 and 1980–4 (BBS, 1978; 1985b; 1986d).[6] The overall annual yield per cultivated hectare of land has risen substantially in Bangladesh. For instance, agricultural value added per net cropped hectare at 1972–3 prices rose from 2714 taka to 3315 taka between 1967–9 and 1982–4, that is, by 22 per cent over a 15-year period.[7]

Cultivated land that was once left fallow for a significant part of each year (for example during the dry season) is now used for crops such as wheat or dry season rice varieties. Even if yields should fall during the original cropping period, the extra cropping in each year in most cases much more than offsets any fall during the original cropping period.[8] This is an important way in which the new varieties have added significantly to agricultural production. A serious shortcoming of the decomposition methods considered in this chapter is their failure to identify such effects. The real contribution of the Green Revolution to increased agricultural output in Bangladesh appears to lie in its contribution to increased productivity of already cultivated land through multiple cropping rather than in its contribution to extension of the area of cultivated land or to yields from single cropping.

3 Measuring the Contribution of Biochemical Technology to Increased Land Productivity in Bangladesh

3.1 INTRODUCTION

The preceding chapter provided empirical evidence about the growth of crop production in Bangladesh and pointed out that the means of increasing agricultural supply in Bangladesh have altered from extension of area cropped to intensified cultivation of land already cultivated following the Green Revolution. The objective of this chapter is to assess quantitatively the importance of biological technology in increasing foodgrain production on cropped land following the Green Revolution. Many writers have claimed that the new biochemical components of the Green Revolution technologies have the greatest impact in raising cropped land. The main purpose is to apply the innovative empirical study by Diwan and Kallianpur (1985) in a new context, namely Bangladesh, to see whether the empirical results conform to those obtained from India and to comment critically on the method employed by Diwan and Kallianpur (1985, 1986).

First of all this chapter briefly reviews relevant literature and, following the methodology employed by Diwan and Kallianpur (1985), specifies the analytical framework adopted. The model is specified empirically in order to estimate the contribution of biological technology to Bangladeshi foodgrain productivity per hectare of cultivated land. The empirical findings are compared with those of the Diwan-Kallianpur (D-K) study for India. In contrast to the D-K study, it is argued that the fertilizer input has made a significant contribution to productivity increase in Bangladeshi foodgrains. This follows from the conclusions reached via reinterpretation of the D-K

51

model and is supported by other evidence. It is contended that the D-K approach is subject to severe limitations and that it can give a misleading view of the contribution to production of the new technology associated with the Green Revolution.

Given an extremely unfavourable land–man ratio, increasing productivity per hectare of cultivated land assumes critical importance. In a recent study of Indian agriculture, Diwan and Kallianpur (1985) (see also Diwan and Kallianpur, 1986) try to quantify the contribution of biological technology (BTC) to agricultural production. Adopting the Hayami-Ruttan (Hayami and Ruttan, 1971; Ruttan *et al.*, 1978) definition of BTC, they introduce a new parameter for BTC and formulate a reduced form capable of quantifying this parameter. Taking fertilizer as a proxy for BTC and utilizing time series data, they claim on the basis of their Indian data that BTC has made a positive contribution to wheat production but not to rice. They conclude that 'the weak impact of BTC on food production, therefore, raises major issues for research and policy' (Diwan and Kallianpur, 1985, p.635).

3.2 CONCEPTUAL ISSUES AND ANALYTICAL FRAMEWORK

Following Solow's seminal work (Solow, 1957), it is generally agreed that increases in the quantities of inputs like labour and capital are not sufficient to explain productivity growth. The explanation of productivity change lies to a considerable extent in technological change. The attention of this chapter is focused on this factor.

Schultz (1964, pp.36–52) hypothesized that traditional farmers may be poor but nevertheless efficient, with the implication that: (1) increased production can result from the extension of the isoquant and not by reallocation of resources, and (2) the extension of the isoquant can take place only when technological change occurs. Given this approach, taken by a group of agricultural economists, neither X-inefficiency nor allocative inefficiency are sources of significant production loss, and the general assumptions of *neoclassical economics* hold.

Technological change is an important factor influencing agricultural productivity growth. In this context 'technological change can be defined as the introduction of new or non-conventional resources into agricultural production as substitutes for the conventional resources. The effect of technological change must be evident as a change in the yield per acre, as a change in the cultivated acreage available, or both' (Yudelman *et al.*, 1970, pp.38–9). It is often a vehicle for the application and substitution of knowledge for material resources or for the substitution of a less expensive resource for a more expensive and scarce one.

Hayami and Ruttan (1971) identified the capacity to develop technology consistent with physio-cultural endowments as the single most important factor accounting for differences in agricultural productivity among countries. Historically, the United States and Japan, starting with entirely different factor endowments, experienced high rates of agricultural growth. In the United States, the expansion of agricultural output resulted primarily from the introduction of technology which facilitated an increase in the area operated per worker. On the other hand, agricultural output growth in Japan was the outcome of the introduction of technology leading to increased yield per cropped hectare in response to fertilizer application, despite a highly unfavourable land–man ratio (cf. Grabowski and Sivan, 1983).

Recent studies have classified agricultural innovations into various categories. Hayami and Ruttan (1971, pp.44–5) and Ruttan *et al.* (1978, p.46) distinguish between biological and mechanical technologies.[1] Traditionally mechanical technology is identified as labour-saving: biological and chemical technology as land-saving. The former facilitates substitution of power and machinery for labour, while the latter is designed to facilitate substitution of labour and/or industrial inputs for land. This distinction is akin to the one between 'landesque' and 'labouresque' capital employed by A.K. Sen (1960, p.91) even though this division is not considered to be watertight. Yudelman *et al.* (1970, p.39) argue that 'land-saving innovations are considered to be reflected solely in an increase in yield and labour saving innovations are considered solely in an increase in the land–man ratio'. Kaneda (1982) makes a similar distinction between labour-saving and land-saving technologies.

In terms of the above, it is clear that land productivity can be increased through the employment of biological technology. In countries where there is little or no scope for extension of cultivation, total agricultural output can only increase if yield per hectare rises. Ruttan *et al.* (1978) define biological technology in terms of an identity which expresses output per worker employed as the product of the yield per hectare and the land–labour ratio. More specifically, Ruttan *et al.* (p.46) point out, increases in output per worker can also be achieved through increased land productivity, but only if the rate of increase in output per hectare exceeds the rate of change in the number of workers employed per hectare. It is useful to refer to these technologies that increase output per hectare as biological technology.

In order to appreciate the quantitative significance of BTC, Diwan and Kallianpur (1985) postulate the following production function:

$$(X_p)^{-\varrho} = (A_a X_a)^{-\varrho} + (A_1 X_1)^{-\varrho} \tag{3.1}$$

where X_p, X_a, X_1 refer to the physical quantities of output, land and labour respectively and A_a and A_1 define land and labour in terms of efficiency or quality. Equation (3.1) is a neoclassical production function involving one parameter of substitution ϱ between land and labour. The elasticity of substitution σ is related to ϱ as follows:

$$\sigma = 1/(1 + \varrho) \tag{3.2}$$

Thus $\sigma \to 1$ only if $\varrho \to 0$. Theoretically this implies that the elasticity of substitution σ can take any value in the range $0 < \sigma < \infty$.

Note that harmonic production function (3.1) is homogenous of degree one and, therefore, exhibits constant returns to scale. This linearly homogenous production function has been employed for two reasons: First, it is the same as that used by Diwan and Kallianpur (1985), and our study involves the application of the D-K analysis in a new context and comments on this. Secondly, empirical studies using farm-level data provide some support for the existence of constant returns to scale. For instance, I. Ahmed (1981, p.50), using Bangladeshi data, found '. . . the returns to scale are not significantly different from constant returns to scale and this finding is equally valid for traditional and high yielding variety rice crops.' Yotopoulos and Nugent (1976, p.70), using Indian data, reported similar evidence. Both of these studies employ Cobb-Douglas production functions, unlike the D-K study. When both *land* and labour are variable, constant returns to scale for production in agriculture does not seem to be an unrealistic assumption.

Employing the conceptual basis embodied in the definition of BTC by Ruttan *et al.* (1978) and the symbolic notations in (3.1), it can be argued that the impact of biological technology can be measured by the difference between the rate of growth of land productivity and the rate of growth of labour–land ratio. Taking the rate of change over time and adopting the approach of Diwan and Kallianpur (1985), one can write

$$(x_a - x_1) > 0 \Rightarrow \text{BTC} \tag{3.3}$$

where $x_a = d/dt(X_p/X)]/(X_p/X_a)$ and $x_1 = d/dt(X_p/X_1)/(X_p/X_1)$. This is reminiscent of Solow's definition (Solow, 1957, p.313) of technological change as the difference between the rate of growth of labour productivity and capital–labour ratio. One needs to be reminded, however, that BTC as defined in (3.3) is a surrogate concept.

From (3.1) the following growth equation can be derived:

$$x_a - x_1 = -S_1x_1 + S_1q_L + S_2a_1 \tag{3.4}$$

where $q_L = [d/dt(A)]/A_a$; $a_1 = [d/dt(A_1)]/A_1$ and S_1 and S_2 respectively represent factor shares of land and labour and $S_1 + S_2 = 1$. The expression on the left-hand side of (3.4) is the same as those of (3.3) and symbolizes BTC which, however, is not precise enough. A rearrangement of (3.4) gives

$$x_a - S_2x_1 = S_1q_L + S_2a_1 \tag{3.5}$$

which is quite general, and the right-hand side consists of terms measured in efficiency units. As $S_2 < 1$, BTC in (3.5), which conforms to Solow's definition, is biased upwards.

The formulation (3.1) above contains qualitative variables A_a and A_1 which are unobservable. In this chapter the primary concern is with the contribution of BTC in relation to land quality. It is assumed that other factors remain the same. The implication of this assumption is that the qualitative variable designated A_1 symbolizing the quality of labour inputs in (3.1) remain constant so that $a_1 = 0$.

BTC as a composite index of agricultural innovations is a function of several component technologies, e.g. HYV seeds, irrigation water, chemical fertilizer and pesticides. Based on the evidence from Bangladesh and Japan, Hayami and Ruttan (1971, pp.82–3) rightly argue that HYVs have a much higher fertilizer response curve than indigenous varieties. M. Khan (1981, pp.156–7) shows a higher fertilizer response curve for irrigated HYVs compared to rainfed HYVs of paddy in Bangladesh. In other words, BTC leads to a shift in fertilizer-response surface and hence can be a catalyst in substituting fertilizer for land. One would expect, therefore, that the process of technological change in agriculture as envisaged by BTC may be captured by fertilizer. To the extent the relationship between other inputs like HYVs, irrigation and agro-chemicals on the one hand and chemical fertilizer on the other is one of complementarity, it can be hypothesized that the efficiency parameter of land, i.e. A_a, depends on fertilizer intensity of land. Recent studies (e.g. Antle, 1983; M. Khan, 1981) have incorporated fertilizer as a separate variable in the production function.

To incorporate the above idea into the framework of this analysis, as is done by Diwan and Kallianpur (1985), the present study assumes a function (g) indicating a relationship between fertilizer consumed per unit of land (F) and the quality of land (A_a) (Diwan and Kallianpur, 1985, p.629) of the following form:

$$q_L = gf \tag{3.6}$$

where $f = d/dt[F]/F$. One could term F as the fertilizer intensity of land. Even though the exact form that g is likely to take has not been specified, *a priori* $g' > 0$. This implies higher fertilizer intensity of land would augment land fertility and hence enhance its quality. However, $g'' < 0$, because of the law of diminishing returns to a variable factor and perhaps because of saturation or bottlenecks in the supply of other complementary inputs. As an approximation, the following simple linear form of (3.6) is proposed:

$$q_L = \gamma f \tag{3.7}$$

with $\gamma > 0$. The coefficient of fertilizer intensity defines the elasticity of land quality with respect to the intensity of fertilizer application. Substituting (3.7) into (5.5) and assuming A_1 to be constant, Equation (3.8) is derived

$$x_a - S_2 x_1 = S_1 f \tag{3.8}$$

where the term on the right-hand side of (3.8) is a measure of BTC which depends on: (1) the rate of growth in fertilizer intensity, (2) factor share of land, and (3) the elasticity of land quality with respect to fertilizer.

One can easily see that the formulations (3.2) and (3.7), i.e. the production function and the BTC function, are structural relations and do not include any economic considerations. To incorporate economic arguments into the framework of analysis, assuming farmers are prepared to pay a fertilizer price equivalent to its marginal product, relationship (3.9) is postulated as follows:

$$d/dt(f_3)/f_3 = d/dt(P_f)/P_f \tag{3.9}$$

where f_3 is the partial of output with respect to fertilizer-intensity and P_f is the price of fertilizer. From (3.1), (3.7) and (3.9), equation (3.10) is obtained:

$$Y = (X_p/X_a) = a_0(P_f)^{a_1}(F)^{a_2} \tag{3.10}$$

$$\text{where } a_0 = A_{10}C \tag{3.11}$$

A_{10} and C are constants of integration

and

$$a_1 = (1/1 + \varrho) \tag{3.12}$$

$$a_2 = (\varrho\gamma + 1)/(1 + \varrho) \tag{3.13}$$

Diwan and Kallianpur (1985, p.630) interpret Equation (3.10) as an explanation of land productivity in terms of fertilizer prices, as well as elasticity of substitution between land and labour on the one hand and the impact of intensity of fertilizer use and land quality on the other. In the present chapter, fertilizer is taken as a proxy for BTC. Hence a sequential causality can be formulated (Hicks, 1980, pp.87–102): BTC extends the frontiers of production by improving land quality which in conjunction with land–labour substitution augment land productivity. The full model can now be presented as follows:

$$(X_p)^{-\varrho} = (A_a X_a)^{-\varrho} + (A_1 X_1)^{-\varrho} \tag{3.1}$$

$$A_\alpha = A_0 Z^\gamma \tag{3.7'}$$

$$Y = (X_p/X_a) = a_0(P_f)^{a_1}(F)^{a_2} \tag{3.10}$$

These three equations contain six variables: output, land, labour, fertilizer, A_a and P_f, of which output, land and A_a are endogenous and labour, fertilizer and P_f are exogenous. Equation (3.10) indirectly contain σ, ϱ and γ which are parameters of interest.

One can note that Equation (3.10) is exactly identified as it satisfies (within the system of equations in the model) both order and rank conditions (Johnston, 1984, pp.454–60). Order condition is satisfied in that at least two variables A_a and X_1 are excluded. The rank condition is that the determinant D be non-singular, where

$$D = \begin{vmatrix} -\varrho & -\varrho \\ 0 & -1 \end{vmatrix}$$

implying that $D = 0$ only if $\varrho = 0$. The rank condition, therefore, is that $\varrho \neq 0$, otherwise the production itself becomes degenerate.

The above model contains only one relation, Equation (3.10), that can be estimated. To estimate Equation (3.10) it is assumed that it is stochastic. Because of the fact that (3.10) is exactly identified, the principle of indirect least squares (ILS) can be applied to estimate reduced-form coefficients by ordinary least squares (OLS) and to compute the structural parameters by an appropriate transformation of the estimated reduced-form coefficients.

As the estimates of the structural parameters ϱ and σ are to be derived from the reduced-form coefficients a_1 and a_2 one should specify the *a priori* expectations about their signs and magnitudes. Obviously a_1, $a_2 \neq 0$. Furthermore, if $\varrho \neq 0$, it follows from (3.12) that $a_1 \neq 1$. It is clear from (3.13) that a_2 must be different from 1. Thus both a_1 and a_2 must be different from 1 as well as 0. One needs to appreciate that this is rather a stringent assumption.

3.3 EMPIRICAL RESULTS

The data presented in Table 3.1 have been used to estimate Equation (3.10). As the required data on all the variables are available only from 1967–8, the analysis of the present study regarding the contribution of BTC is confined to the period 1967–8 to 1982–3. As indicated earlier, the present analysis considers two types of yields (yield per gross and net cropped hectare) for rice (GRYLD and NRYLD) and for foodgrains (GFDYLD and NFDYLD) as a whole.

Because wheat is cultivated only once a year, there is no distinction between gross and net area planted to wheat, and between its yield per gross and net cropped hectare (WHYLD). In order to provide some indication of the spread of the new technology over the years (rather than any two points in time), percentage areas of all food-grains, rice and wheat under HYV cultivation (respectively designated as PRFDHA, PRICEHA and PRWHHA) and percentage area under irrigation in foodgrains (IPCFOOD) are also presented in Table 3.1.

The results of the estimates are presented in Table 3.2. It can be seen that the present model has a reasonably high explanatory power and the estimated coefficients are significant at acceptable levels of probability. All the sample Durbin-Watson (DW) statistic values are higher than the upper limit of its tabulated value of 1.252 for 16 observations and two explanatory variables. Therefore, there does not appear to be any problem associated with *positive* first-order autocorrelation. Since the DW values are less than or close to 2, it is not necessary to test for *negative* first-order autocorrelation (J. Johnston, 1984, pp.315–16). Because F measures fertilizer intensity of land, one would *a priori* expect yield to be positively correlated with it and hence $a_2 > 0$. On the other hand, PFR or PFW measures price of fertilizer relative to the price of paddy or wheat. A rise in fertilizer price relative to that of either crop would influence profitability and is likely to have a disincentive effect on fertilizer use for their production. That is, the correlation between PFR or PFW and yield is expected to be negative and $a_1 < 0$. In the light of the above, the estimates of a_2 seem to be consistent with *a priori* expectations as are those of a_1. This is supported by a more recent study (M. Hossain, 1985). Using pooled time series and district-level cross-sectional data, Hossain finds (p.58) that a rise in fertilizer price would lead to a fall in consumption. However, 'a decline in consumption in response to higher fertilizer prices could be compensated by an increase in the area under modern irrigation and/or by a change in cropping pattern in favour of high yielding varieties. The negative impact could also be compensated by an increase in paddy prices'. Hossain further suggests that the expansion of areas under irrigation and HYVs are the most important factors explaining increased fertilizer use in the country.

On the basis of the estimates of a_1 and a_2 presented in Table 3.2, the estimates of the structural parameters ϱ and γ can be derived

Table 3.1 Intensity of modern input use, fertilizer–paddy and fertilizer–wheat price ratios and yields of different foodgrains per gross and net cropped hectare: Bangladesh, 1967–8 to 1982–3

Year	PRFDHA	PRICEHA	PRWHHA	IPCFOOD	PFR	PFW	KGHA	KGNHA	NFDYLD	NRYLD	GFDYLD	GRYLD	WHYLD
1967	0.686	0.679	1.562	NA	0.603	0.671	7.898	11.418	1671.18	1675.17	1155.95	1158.71	804.36
1968	1.648	1.581	7.241	NA	0.504	0.561	8.499	12.237	1698.40	1702.98	1179.58	1182.75	915.49
1969	2.618	2.558	7.770	8.725	0.525	0.584	9.908	14.951	1753.70	1756.39	1162.13	1163.92	1008.60
1970	4.720	4.642	10.932	10.174	0.567	0.631	11.771	16.738	1597.29	1598.51	1123.31	1124.18	1055.47
1971	6.779	6.710	11.783	10.304	0.411	0.457	10.170	14.063	1477.22	1477.05	1068.28	1068.16	1077.45
1972	11.140	11.056	17.845	10.863	0.484	0.602	15.459	21.541	1455.28	1460.21	1044.39	1047.93	760.82
1973	15.772	15.675	23.606	11.583	0.477	0.368	14.890	20.886	1685.77	1691.11	1201.82	1205.63	897.28
1974	14.946	14.803	26.045	13.103	0.362	0.323	11.371	15.839	1608.20	1612.27	1154.55	1157.47	928.40
1975	15.651	15.024	58.760	12.100	0.637	0.806	17.626	24.955	1753.70	1749.25	1238.68	1235.54	1454.99
1976	13.904	12.949	72.911	10.474	0.831	0.700	20.343	28.834	1695.53	1685.79	1196.22	1189.35	1620.84
1977	16.069	14.843	83.298	12.145	0.702	0.678	26.941	40.610	2006.47	1992.51	1331.10	1321.85	1838.69
1978	18.694	16.922	89.144	12.225	0.703	0.796	25.804	39.505	2009.38	1988.07	1312.48	1298.56	1865.77
1979	22.735	19.662	94.771	13.229	0.780	0.796	28.794	44.107	1961.40	1920.98	1280.44	1254.06	1898.87
1980	25.368	21.281	96.646	13.334	0.789	0.796	31.955	49.117	2111.48	2069.72	1373.70	1346.53	1847.38
1981	25.845	22.227	96.715	13.904	0.945	0.845	30.171	46.421	2042.96	2004.97	1327.79	1303.10	1811.43
1982	28.161	24.837	95.947	14.885	1.003	0.889	37.945	58.685	2132.26	2076.82	1378.69	1342.85	2109.55

Notes: 1967 refers to 1967–8 (July 1967 to June 1968), etc. PRFDHA, PRICEHA, PRWHHA represent percentage area of all foodgrains, rice and wheat under HYV cultivation. KGHA and KGNHA are fertizer (kg of nutrients) applied per hectare of gross and net cropped area. IPCFOOD represents percentage of foodgrain area irrigated. PFR and PFW are fertilizer–paddy and fertilizer–wheat price ratios respectively. NFDYLD and GFDYLD respectively refer to foodgrain yield per net and gross cropped hectare. NRYLD and GRYLD respectively refer to yield per net and gross hectare cropped with rice. WHYLD is yield of wheat per hectare. NA means not available.

Source: Fertilizer prices are weighted averages of those of urea, TSP (triple super phosphate) and MP (muriate of potash). Fertilizer prices up to 1973–4 and paddy prices up to 1971–2 are from M. Khan (1981, p.161). Prices of fertilizer, paddy and wheat are based on information contained in USAID (1982, p.70) and BBS (1984b, p.525; 1985a, p.792–8). Wheat prices prior to 1972–3 are derived assuming a wheat–paddy price ratio of 0.90. The ratio in 1972–3 was marginally greater than 0.90. Data on all other variables are based on the information from BBS (1976, pp.1–2, 4–10, 12–5, 26–9; 1979, pp.162, 168–71, 212; 1980a, pp.20–5, 30–1, 33–4, 36–7, 46–52; 1982, pp.206, 209, 213, 232, 235–8, 240–1; 1984a, pp.31, 33, 39, 42; 1984b, pp.225, 249–52, 255); BRRI (1977, p.89); EPBS (1969, pp.40–1, 120); World Bank (1982, Tables 2.5–6).

Table 3.2 Estimates of reduced form coefficients in Equation 3.10: Bangladesh, 1967–8 to 1982–3

Crop	Variable	Coefficients			R^2	DW
		a_0	a_1	a_2		
Foodgrains	net yield	7.1918	-0.1648^c	0.1151^b	0.744	1.62
			(1.620)	(2.100)		
Foodgrains	gross yield	6.9136	-0.1094^c	0.08531^b	0.718	1.85
			(1.520)	(2.080)		
Rice	net yield	7.2183	-0.1534^c	0.1033^b	0.715	1.63
			(1.530)	(1.920)		
	gross yield	6.9341	-0.0964^c	0.0738^b	0.667	1.85
			(1.370)	(1.840)		
Wheat	yield	5.8071	-0.2865^c	0.5233^a	0.836	2.01
			(1.610)	(5.190)		

Figures in parentheses are *t*-values;
DW is Durbin-Watson statistic;
[a] Significant at 1 per cent level;
[b] Significant at 5 per cent level;
[c] Significant at 10 per cent level.

Table 3.3 Estimates of structural parameters: Bangladesh, 1967–8 to 1982–3

Crop	Variable	σ	ϱ	γ
Foodgrains	net yield	-0.16476	-7.0694	0.2402
	gross yield	-0.10939	-10.1416	0.1755
Rice	net yield	-0.15343	-7.5176	0.1976
	gross yield	-0.09640	-11.3734	0.1148
Wheat	yield	-0.28653	-4.4900	0.6295

from Equations (3.12) and (3.13) as follows:

$$\varrho = (1/a_1 - 1) \text{ and } \sigma = a_1 \tag{3.14}$$

and

$$\gamma = [(1/\varrho)(a_2/a_1 - 1)] \tag{3.15}$$

The results of the estimates of the present study for σ, ϱ and γ are presented in Table 3.3.

At this stage it is also appropriate to comment on the *a priori* expectations about the sign or magnitude of γ. A commonsense view is that biological technology makes a positive contribution to the growth of crop yields. Therefore, on *a priori* grounds $\gamma > 0$. Diwan and Kallianpur (1985, p.630), however, argue 'it is quite possible that because of misapplication of this technology or because of its side effects, $\gamma >$ or < 0'. Misapplication or side effects can result from unavailability of other complementary inputs in time, as well as lack of balance in their application. About the magnitude of γ, Diwan and Kallianpur (1985, p.635) suggest that 'on the basis of conventional wisdom, one would think that γ should be close to 1'. However, this may not be plausible with the Diwan-Kallianpur model. It can be easily proved from Equations (3.14) and (3.15) that as $\gamma \rightarrow 1$, $a_2 \rightarrow 1$. This violates the basic assumption that a_2 has to be *significantly different* from unity (Diwan and Kallianpur, 1985, p.633).

On the basis of the above discussion, the γ values for Bangladeshi wheat and rice (see Table 3.3) are consistent with *a priori* expectations. Those for foodgrains as a whole are also plausible. However, one needs to be cautious about interpreting these values. First of all these values are not very high. Secondly, not all of the reduced form coefficients from which these values are derived are significant at 5 or 1 per cent levels of probability. Thirdly, the overall quality of the data is not beyond question (see, for example, Alamgir 1980; Boyce 1985, 1987a; Alauddin and Tisdell 1987). Under the circumstances, one could at best say from the above results that BTC does not seem to have made any significant contribution to increased yields of either rice or foodgrains as a whole.

The above findings raise several questions: How do the results of this study compare with those of Diwan and Kallianpur (1985)? How can one explain the different results for wheat and rice?

The results of this study are different from those of Diwan and Kallianpur in two respects. First, the γ value for wheat in the present analysis is much higher than theirs (0.6295 as against 0.06). Secondly, the impact of BTC on rice yield is positive in the context of Bangladesh compared to its negative value in the Indian context (0.1976 as against -0.17). Furthermore, the elasticity of substitution values in the present study are systematically lower than those reported by Diwan and Kallianpur (p.633) for India. One reason for the difference in the values of the parameters of these two studies could be the nature of the data used. Diwan and Kallianpur have used absolute quantities of nitrogenous fertilizer and its price (p.632). They have ignored the use of other important fertilizers like triple super phosphate (TSP) and muriate of potash (MP). By taking absolute quantity rather than nutrient content they may have overestimated the use of fertilizer.[2] Also, by ignoring the use of TSP and MP, they might have introduced bias in the price of fertilizer and hence in the fertilizer–rice or fertilizer–wheat price ratio. The present study, on the other hand, has used the fertilizer variable in terms of the nutrient contents of all the three major fertilizers, i.e., urea, TSP and MP which in recent years account for almost all of the chemical fertilizers used in Bangladesh (BBS, 1984b, p.225).

A number of factors account for the contrasting impact of BTC on wheat and rice yields in Bangladesh. First, consider the case of wheat. Wheat is cultivated only on 3 per cent of the gross area cropped with foodgrains. Virtually all the wheat area is under HYV cultivation. It is grown only during one season, namely the *rabi* (dry) season and more than a third of the wheat area is supplemented with controlled irrigation. As mentioned earlier in this chapter, HYVs are more responsive to fertilizer application compared with the traditional varieties. They are even more so when irrigated. Furthermore, the cultivation of wheat is confined to selected districts. The districts of Khulna, Barisal, Patuakhali, Chittagong, Chittagong Hill Tracts, Sylhet and Rangpur grow little or no wheat. The range of climatic and environmental variabilities is much narrower in the case of wheat. Also wheat as a crop is much less location specific, that is, adaptable to a wide range of variable ecological conditions. As Evenson and Kislev (1973, p.1312) put it, 'Mexican wheat varieties are grown successfully from North Africa to India'.

As for rice, it is cultivated on all types of crop lands throughout the year. Thus the range of environmental and climatic variabilities is much wider in case of rice than it is for wheat. HYVs of rice are not planted on all crop lands. In recent years about a quarter of the gross area cropped with rice is under HYVs, well over 60 per cent of which is grown primarily under rainfed conditions supplemented with very little or no controlled irrigation. Even though nearly 80 per cent of the total dry season foodgrain area is irrigated and 70 per cent is under HYV cultivation, overall foodgrain area irrigated is only around 14 per cent of the gross foodgrain area, and for rice it is even smaller, albeit marginally. Furthermore, rice as a crop is much more location specific (see, for example, Farmer 1979). Most of the HYVs of rice that are currently in use in Bangladesh record high yield potentials under controlled environmental conditions. Once the crop varieties are released for cultivation in farmers' fields, however, it is difficult to duplicate these conditions. Some varieties are more susceptible than others to moisture stress, temperature change and to length of photoperiod, for instance. There is, therefore, bound to be divergence between potential and actual yields. The likely factors that account for such divergence include inadequate and inappropriate application of fertilizers (IFDC 1982); lack of proper plant protection measures (BRRI 1977); differences between recommended practices and those actually followed by farmers; and inadequate irrigation (IRRI, 1974).

Differences in: (1) the range of environmental and climatic variabilities confronted by rice and wheat; (2) their respective degree of location specificity and (3) per cent area under irrigation and HYVs for the two cereals seem to account for the differential impact of BTC on their yields.

In order to analyse their data further, Diwan and Kallianpur (1985) estimated Cobb-Douglas functions involving land and fertilizer as inputs. The Cobb-Douglas estimates seemed to be consistent with the 'magnitude and direction of the γ values' (p.634) for wheat and rice. However, one is uncertain about the logic of including time as an explanatory variable in the Cobb-Douglas function (p.634). Even though detailed data are not provided, it seems likely (from the information contained in Table 2, p.632) that the two variables, time and fertilizer, would be highly correlated. Their inclusion as explanatory variables in the same equation is likely to lead to the problems of multicollinearity reducing the significance of the coefficients of either variable.

A closer examination of Tables 3 and 5 of the D-K study indicates, however, that the coefficients of fertilizer for rice have opposite signs (compare a_2 in Table 2 and b_2 in Table 5). Only when the γ value and the b_2 value (Tables 3 and 5, pp.633–4) for rice are examined does agreement seem to result in terms of direction of the parameters of BTC. Why this conflict? The answer is unclear. Diwan and Kallianpur themselves do not rule out the possibility of misspecification. Whether misspecification has really occurred cannot be readily detected because of the apparent but somewhat dubious similarity between the γ values in Table 3.2 and b_2 values in Table 5 of the D-K study.

To pursue the point further, several equations involving fertilizer, land area under cultivation and yields of foodgrains as a whole and those of rice were estimated. The results are set out in Table 3.4. It can be seen from the table that all the equations show positive

Table 3.4   Estimates of impact on foodgrain yields of fertilizer application: Bangladesh, 1967–8 to 1982–3

Crop	Variable	Coefficients				R^2	DW
		Intercept	Fertiliser	Gross area	Net area		
Rice	gross	−5.062	0.0531[b] (2.110)	1.303[a] (3.670)	–	0.813	2.344
	gross	6.763	0.1179[a] (4.770)	–	–	0.620	1.348
	net	12.492	0.1659[a] (4.940)	−0.628 (0.770)	–	0.678	1.190[c]
	net	6.927	0.1708[a] (5.260)	–	–	0.664	1.018[d]
Foodgrains	gross	−5.247	0.0353 (1.180)	1.326[a] (4.130)	–	0.856	2.294
	gross	6.719	0.1353[a] (5.300)	–	–	0.667	1.265
	net	11.229	0.1960[a] (5.090)	–	−0.4941 (0.480)	0.698	1.504[c]
	net	6.879	0.1876[a] (5.620)	–	–	0.693	0.963[d]

(All variables are in natural logarithm; figures in parentheses are t-values);
[a] Significant at 1 per cent level;
[b] Significant at 5 per cent level.
[c] Tested for first and second-order autocorrelation. But the t-values turned out to be insignificant. Consider the following two relationships for net yield of rice and that of foodgrains respectively:

$$e_t = 0.34410e_{t-1} - 0.12980e_{t-2} \text{ and } e_t = 0.40010e_{t-1} - 0.11650e_{t-2}$$
$$(1.190) \qquad\qquad (0.470) \qquad\qquad\quad (1.380) \qquad\qquad (0.420)$$

[d] Tested for first and second-order autocorrelation. However, the coefficients lacked statistical significance. This can be seen from the following relationships for net yield of foodgrains and rice respectively:

$$e_t = 0.08260e_{t-1} - 0.02890e_{t-2} \text{ and } e_t = 0.41120e_{t-1} - 0.10430e_{t-2}$$
$$(0.290) \qquad\qquad (0.110) \qquad\qquad\quad (1.430) \qquad\qquad (0.380)$$

contribution of fertilizer. One needs to point out that because of high collinearity between fertilizer and gross area (0.7456) a separate equation with fertilizer as the only explanatory variable was estimated. The results are satisfactory from the statistical point of view. High collinearity between gross area under foodgrain cultivation and fertilizer (0.8533) also affects the statistical qualities of the coefficients of both the explanatory variables. Because foodgrain yield per gross cropped hectare has a stronger correlation (0.9170) with gross area cropped with foodgrains than it has with fertilizer (0.8486), the coefficient of the fertilizer variable turns out to be statistically insignificant in the equation in which both of these factor, are included as explanatory variables. When the impact of fertilizer is examined separately the coefficient seems to be much greater in magnitude and the statistical quality improves markedly. Fertilizer application seems to have stronger impact on net yield per hectare than yield per gross cropped hectare.

It is therefore possible that fertilizer makes a substantial contribution to land productivity. This is consistent with the findings of some recent studies (see, for example, Alauddin and Mujeri 1986a; M. Khan 1981; Mahabab Hossain, 1985). These studies, using district-level cross-sectional data and farm-level data, report that chemical fertilizers have a significantly positive impact on foodgrain yields. Yamada and Ruttan (1980, pp.445–6, 455) use cross-country time series data for the 1960–70 period and examine the relationship between yield per hectare and intensity of fertilizer use. Their findings suggest a positive relationship between these two variables. However, the relationship seems to be stronger for the countries already using relatively high levels of fertilizer per hectare. Thus it seems possible that a country like Japan with a highly developed network of agricultural research and extension services, and a relatively assured supply of complementary inputs, can provide an environment whereby crop varieties can become more responsive to increased usage of chemical fertilizers than a country like India or Bangladesh. The relatively weaker forces of diffusion as well as poor quality of management seem to be responsible for inter-country differences in the impact of fertilizer on land productivity.

3.4 LIMITATIONS OF THE D-K APPROACH: SOME FURTHER OBSERVATIONS

Apart from the problems mentioned above, the D-K approach suffers from further limitations. These include problems associated with identification, and limitations imposed by the use of aggregate data.

Consider the identification problem first. The introduction of the new seed-fertilizer-irrigation technology is liable to shift the land quality–fertilizer function (that is, percentage rate of change, see Equation (3.6)). In other words, the parameter γ itself might change over time if the form in Equation (3.7) is adopted. But the D-K formulation of this function (Equation (3.7) above) does not take this factor into account. Therefore it involves the problem of identification of 'true' parameters.

This is illustrated with the aid of Figures 3.1 and 3.2. Consider a simple linear relationship between percentage change in fertilizer application per hectare (f) and that in land quality (q_L) as assumed by D-K and specified in Equation (3.7). Let OA be the curve representing the relationship between f and q_L before or at the very early stages of the introduction of new HYV technology. Assume that the true relationship shifts to OB in the later period. If one assumes that fertilizer intensity is lower in the later period compared to the earlier (as it might be the case because of smaller initial values), a line such as CD or EF might be estimated from observed values. Neither line is the true response function. The first case occurs when values in the neighbourhood of D are observed in the initial period and values in the vicinity of C are observed in the subsequent period. The second case occurs when values near F are observed in the first period and those around E are observed in the subsequent period. While CD is weakly positive, EF is negative. One clearly gets an underestimation of the responsiveness coefficient γ in both situations, and this is greatly underestimated for the newest technology.

Figure 3.2 depicts a different situation. Here autonomous improvement in land quality independent of fertilizer application is possible (for example, because of agricultural innovations). Suppose that the initial response function is represented by AB and the subsequent one by CD. The intersection point of these functions is represented by P. Assume f to be larger in the later period than in the earlier. Then if the initial observations fall to the left of P around a point on AB and the subsequent observations tend to fall on CD, the estimated response function underestimates the responsiveness coefficient γ in the ultimate period of BTC. For example, if initial observations are around E on AB (the old response function) and subsequent observations are in the neighbourhood of H on CD (the new response function), a relationship such as EH (showing negative response) might be estimated. *Mutatis mutandis*, EF provides another example of a misspecified response function which underestimates responsiveness in the later stages of BTC.

If initial observations tend to fall to the right of P in Figure 3.2, estimated response functions will tend to show smaller slopes than the true one after BTC if the per cent change in fertilizer application per hectare declines compared to the initial one. For example, if

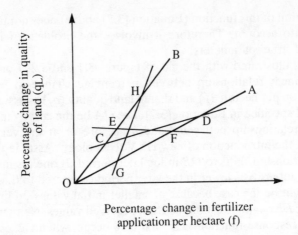

Figure 3.1   The D-K model underestimates the impact of fertilizer application on induced improvement in land quality

points in the neighbourhood of *J* are first observed and subsequently in the vicinity of *F* the line *FJ* may be estimated.

However, one cannot rule out the possibility of upward bias of estimated γ depending upon the nature of observations. For instance, in Figure 3.1 an estimated line such as *FJ* could emerge. This requires observations depending on the percentage change in fertilizer application being larger than the earlier one. In Figure 3.2 reverse bias can also be generated in a similar way to the right of the intersection point *P*.

From the above analysis it is clear that spurious responsiveness estimates can result because of omission of 'shift function', i.e., of the variables that cause the land quality response function to shift (see Koutsoyiannis, 1978, p.348). There may be an identification problem whereby neither the new nor the old function is estimated. The estimated relationship can, in this situation, measure a 'mongrel' function which is a combination of both the new and the old response functions (Tisdell, 1972, p.352). By ignoring the possibility of shifts, this may have happened in the case of the D-K model. Identification problems might explain why the estimates of γ values are smaller than expected *a priori*, and in some cases even negative.

One further complication is that the D-K model assumes full equilibrium. The crucial question is: What if the observations are disequilibrium ones as is most likely the case? Disequilibrium is likely

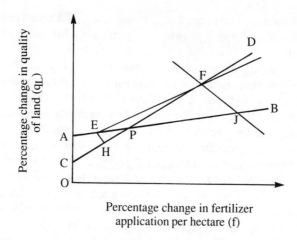

Figure 3.2   An illustration of a misspecified response function
underestimating responsiveness in the later stages of fertilizer application

to aggravate the problem of identification. While there is dispute
about this matter, there is some evidence that farmers in LDCs fail to
maximize profit and may be less than fully efficient in their use of
technical knowledge (A.K. Sen, 1966; Wharton, 1969; Yotopoulos
and Lau, 1973).

Diwan and Kallianpur use highly aggregated data. There have
been a number of biological technical changes in the period considered by them. The effect of these on productivity and 'land quality'
is influenced by the nature and rate of diffusion. Apparent response
will be less where diffusion is slow. There is need to take specific
account of this. It is difficult to have confidence in the D-K results
when experimental and cross-sectional data indicate considerable
response to fertilizer application for new crop varieties (M. Khan,
1981; IFDC, 1982). This seems justified even after taking account of
limitations of experimental and cross-sectional data (De Silva and
Tisdell, 1985).

3.5   CONCLUDING REMARKS

In this chapter, the Diwan-Kallianpur model has been applied to
Bangladeshi data for the first time. While the results for Bangladesh
seem more plausible *a priori* than the D-K estimates for India,

estimated γ coefficients are low. Diwan and Kallianpur (1985; 1986) found lower and more negative γ coefficients for India than they expected.

At first sight, these results suggest that biological inputs have played a much smaller role in increasing food productivity and, in particular, innovations of the Green Revolution have been less significant in this regard than has been commonly supposed. While this conclusion may be warranted, there are grounds for scepticism. In particular, important problems of identification arise when the D-K method is used and aggregation problems need to be overcome. While the approach of Diwan and Kallianpur is innovative and makes use of limited data, more evidence is needed before one can conclude 'from this analysis that contribution of BTC, more specifically fertilizers, to foodgrain production . . . , is quite low' (Diwan and Kallianpur, 1985, p.627). At this stage there is indeed insufficient evidence to reject the opposite hypothesis, namely that the contribution of biochemical technologies to increased foodgrain production has been significant at least in the short run. In the long run, however, as discussed later in this book, the use of such technologies may create difficulties for sustaining yields because of their possible adverse ecological impacts.

4 Regional Variations in Growth Patterns of Bangladeshi Foodgrain Production and Yield

4.1 INTRODUCTION

The last two chapters have examined for Bangladesh growth in overall crop output and productivity per unit area with particular emphasis on foodgrains (rice and wheat). The purpose of this chapter is to concentrate on deriving growth rates of foodgrain production and yield at a more disaggregated level, that is by districts, and use the results for comparative purposes. Recent studies of Bangladeshi agriculture (Alauddin and Mujeri, 1986a; 1986b; Chaudhury, 1981b; M. Hossain, 1980) have identified discernible regional patterns in the pace of growth of agricultural crop output in Bangladesh.

The first two studies mentioned employed disaggregated analysis and estimated growth rates for 17 districts (see Figure 7.1) on the basis of changes between two end-points. For instance, Chaudhury (1981b), taking 19 crops, derived district-level growth rates between 1961–4 (average for the years 1961–2, 1962–3 and 1963–4) and 1974–7 (average for the years 1974–5, 1975–6 and 1976–7). Alauddin and Mujeri (1986a) derived growth rates in aggregate crop output (involving 13 major crops) between 1967–9 (average for the years 1967–8, 1968–9 and 1969–70) and 1978–80 (average for the years (1978–9, 1979–80 and 1980–1). M. Hossain (1980) estimated growth rates in foodgrains by fitting semi-log trends for the 1964–5 to 1977–8 period. All the three studies have one thing in common. They do not make inter-temporal comparisons (e.g. pre- and post-Green Revolution periods) of the growth rates of overall crop or foodgrain output for various districts, in as much as they concentrate broadly on the period of the new agricultural technology.

The present chapter investigates growth in Bangladeshi foodgrain production, employing district-level data, but overcoming the limitations of the previous studies. It uses time series data of a longer duration and makes inter-temporal comparison of growth rates employing a kinked exponential model (see Boyce, 1986). In this chapter we have considered 17 districts which were used in Bangladesh until the later part of the 1960s. At present, Bangladesh is divided into more than 60 administrative districts.

In this chapter growth rates in foodgrain production and yield are estimated by districts. Growth rates in foodgrain production and yield are estimated first of all for each district for the period prior to the Green Revolution (1947–8 to 1968–9) and the period following it (1969–70 to 1984–5).[1] This is followed by an analysis of inter-district growth rates for foodgrain production and yield in the period of the new technology. Differences between growth rates are noted, and factors which may explain differences in regional patterns of growth are identified. We also consider whether growth rates in foodgrain production and yield have become more divergent following the introduction of the new technology. If they have, this can become a source of increasing disparity of farm incomes between districts (a source of greater income inequality between regions) and a cause for concern. Furthermore, the results could be at odds with the view that economic growth tends to result in either a convergence of rates of growth or greater equality of income by regions. But we must not prejudge the matter.

4.2 OBSERVED PATTERN OF GROWTH IN FOODGRAIN PRODUCTION AND YIELD FOR BANGLADESH DISTRICTS, 1947–68 AND 1969–84

On the basis of time cut-off points outlined above, the one-kink exponential model presented as Equation (2.3) in Chapter 2 and reproduced here as Equation (4.1), growth rates in foodgrain production and yield are estimated for each district.

$$1nY_t = \alpha_1 + \beta_1(D_1t + D_2k) + \beta_2(D_2t - D_2k) + e_t \qquad (4.1)$$

A straightforward application of the OLS technique yields estimates of β_1 and β_2 which give exponential growth rates corresponding to the two sub-periods. Thus the relation $\beta_1 \neq \beta_2$ implies the presence of a kink at the point k between the two sub-periods. In the present chapter it is postulated that the kink occurs at 1968–9, which marks the end of the pre-Green Revolution period.

The estimated growth rates of foodgrain production and yield per hectare for Bangladesh and its districts for the two sub-periods are set out in Table 4.1. For Bangladesh as a whole, foodgrain production has increased at a higher rate in Period 2 (2.60 per cent during 1969–84 compared to 2.12 per cent during 1947–68). Foodgrain yield per gross cropped hectare has also recorded a higher rate of growth during Period 2 (1.67 per cent) as compared to Period 1 (1.11 per cent).

The picture of overall foodgrain output growth for Bangladesh as a whole conceals important inter-district differences in output performance. The information presented in Table 4.1 suggests that the growth rate of crop output has been lower than the national average in ten districts in the pre-Green Revolution period and two districts (Faridpur and Khulna) experienced retrogression. The production growth rate was higher than 3 per cent in only in one district, namely, Chittagong. As for yield in Period 1, eight districts had growth rates higher than the national rate and only in one district, namely Chittagong, did yield grow at an annual rate of more than 2 per cent. Faridpur recorded negative growth in yield, while rates of growth of yields in the districts of Barisal, Pabna, Khulna, Jessore, Kushtia and Dinajpur were very sluggish.

There is a significant change in growth rates in yield as well as production in the post-Green Revolution period. Eight districts show production growth rates higher than the one for Bangladesh as a whole, while in twelve districts yields grew at rates faster than that for Bangladesh. In five districts, namely Bogra, Kushtia, Pabna, Dinajpur and Mymensingh, foodgrain production increased at an annual rate of more than 3 per cent, while for seven districts production grew at annual rates between 2.3 and 3.0 per cent. Yields per gross cropped hectare have followed a similar pattern of growth. Furthermore, it can be observed that no district registers negative production growth rates in Period 2 whereas two do in Period 1. As for yields, most districts registered negative (but near zero) growth rates in Period 1 but only one district did so in Period 2. Positive growth both of production and yields were much more common by districts in Period 2 than in Period 1.

On the basis of growth rates in production and yield, four groups of districts can be identified:

Production: Group 1 (above 3.0 per cent); Group 2 (2.5 to 3.0 per cent); Group 3 (1.5 to 2.5 per cent); Group 4 (below 1.5 per cent).
Yield: Group 1 (above 2.0 per cent); Group 2 (1.5 to 2.0 per cent);

Table 4.1 Growth rates of foodgrain production and yield, using one-kink exponential model: Bangladesh districts, Period 1 (1947–8 to 1968–9) and Period 2 (1969–70 to 1984–5)

District	Intercept	Period 1		Period 2		R^2	F-ratio
		Growth rate (%)	t-value	Growth rate (%)	t-value		
(a) Production ('000 metric tons)							
Dhaka	6.0005	1.90[a]	4.11	2.51[a]	3.25	0.6524	32.851[a]
Mymensingh	6.8063	2.86[a]	11.38	3.46[a]	8.23	0.9303	233.727[a]
Faridpur	6.1907	-0.68[c]	1.48	1.92[a]	2.52	0.1531	3.163[d]
Chittagong	5.6084	3.37[a]	7.39	2.92[a]	3.83	0.8169	78.087[a]
Chittagong HT	3.9514	2.72[a]	6.28	0.69[d]	0.96	0.6897	34.449[a]
Noakhali	5.4874	3.47[a]	10.63	2.57[a]	4.71	0.8940	147.627[a]
Comilla	6.2464	1.96[a]	6.26	2.54[a]	4.85	0.8106	74.896[a]
Sylhet	6.3439	2.72[a]	7.19	0.44[d]	0.70	0.7168	44.283[a]
Rajshahi	6.1019	2.06[a]	4.75	1.86[a]	2.57	0.6543	33.125[a]
Dinajpur	5.8592	1.72[a]	2.75	3.53[a]	3.15	0.5389	20.451[a]
Bogra	5.5219	1.98[a]	4.80	4.43[a]	6.41	0.8136	76.400[a]
Rangpur	6.1385	3.49[a]	12.51	2.61[a]	5.59	0.9215	205.439[a]
Pabna	5.5294	0.95[b]	2.23	3.96[a]	5.54	0.6859	38.214[a]
Khulna	6.1590	-0.29[d]	0.42	2.85[a]	2.46	0.1793	3.824[b]
Kushtia	5.1645	0.54[d]	1.13	4.27[a]	5.30	0.6126	27.668[a]
Jessore	5.8411	1.51[a]	4.43	2.44[a]	4.29	0.7244	46.003[a]
Barisal	6.4909	1.66[a]	4.48	1.36[b]	2.20	0.6136	27.784[a]
BANGLADESH	8.8504	2.12[a]	10.48	2.60[a]	6.76	0.9076	171.989[a]

(b) Yield (metric tons/gross cropped ha)

Dhaka	-0.1086	1.13[a]	4.46	2.05[a]	4.83	0.7489	52.189[a]
Mymensingh	-0.1480	1.33[a]	7.47	1.84[a]	6.18	0.8657	112.825[a]
Faridpur	-0.1273	-0.65[b]	2.07	1.94[a]	3.67	0.2786	6.757[a]
Chittagong	-0.1605	2.65[a]	7.94	2.15[a]	3.85	0.8319	86.633[a]
Chittagong HT	0.0004	1.03[a]	3.16	2.17[a]	4.04	0.6575	29.760[a]
Noakhali	-0.2611	1.72[a]	5.45	2.04[a]	3.87	0.7513	52.865[a]
Comilla	-0.1663	1.65[a]	7.40	1.78[a]	4.77	0.8381	90.563[a]
Sylhet	-0.1208	1.77[a]	9.35	-0.08[d]	0.26	0.7864	64.438[a]
Rajshahi	-0.1699	1.01[a]	3.49	1.43[a]	2.97	0.5909	25.279[a]
Dinajpur	-0.0557	0.64[b]	1.94	1.72[a]	2.92	0.4446	14.009[a]
Bogra	-0.1240	1.15[a]	3.58	2.58[a]	4.82	0.7102	42.885[a]
Rangpur	-0.2084	1.50[a]	6.30	1.41[a]	3.54	0.7733	59.692[a]
Pabna	-0.1753	0.21[d]	0.64	2.39[a]	4.28	0.4886	16.717[a]
Khulna	-0.0080	0.34[d]	0.72	1.74[b]	2.22	0.2411	5.561[a]
Kushtia	-0.2058	0.45[d]	1.24	3.24[a]	5.33	0.6218	28.773[a]
Jessore	-0.0810	0.38[d]	1.29	1.46[a]	2.94	0.3906	11.218[a]
Barisal	0.0032	0.18[d]	0.57	1.16[b]	2.15	0.2182	4.884[b]
BANGLADESH	-0.1163	1.11[a]	7.23	1.67[a]	5.71	0.8459	96.071[a]

[a] Significant at 1 per cent level;
[b] Significant at 5 per cent level;
[c] Significant at 10 per cent level;
[d] Not significant.

Source: Based on district-level data from sources mentioned in Table 2.1.

Table 4.2   Groups of districts identified by growth rates in foodgrain production and yield per gross cropped hectare: Period 1 (1947–68) and Period 2 (1969–84)

Group	District
	Production (Period 1)
Group 1 (above 3%)	Rangpur, Noakhali, Chittagong.
Group 2 (2.5 to 3%)	Mymensingh, Sylhet, Chittagong Hill Tracts.
Group 3 (1.5 to 2.5%)	Rajshahi, Bogra, Comilla, Dhaka, Dinajpur, Barisal, Jessore.
Group 4 (below 1.5%)	Pabna, Kushtia, Khulna, Faridpur.
	Production (Period 2)
Group 1 (above 3%)	Bogra, Kushtia, Pabna, Dinajpur, Mymensingh.
Group 2 (2.5 to 3%)	Chittagong, Khulna, Rangpur, Noakhali, Comilla, Dhaka.
Group 3 (1.5 to 2.5%)	Jessore, Faridpur, Rajshahi.
Group 4 (below 1.5%)	Barisal, Chittagong Hill Tracts, Sylhet.
	Yield (Period 1)
Group 1 (above 2%)	Chittagong.
Group 2 (1.5 to 2%)	Sylhet, Noakhali, Comilla.
Group 3 (1 to 1.5%)	Rangpur, Mymensingh, Bogra, Dhaka, Chittagong Hill Tracts, Rajshahi.
Group 4 (below 1%)	Dinajpur, Kushtia, Jessore, Khulna, Pabna, Barisal, Faridpur.
	Yield (Period 2)
Group 1 (above 2%)	Kushtia, Bogra, Pabna, Chittagong Hill Tracts, Chittagong, Dhaka, Noakhali.
Group 2 (1.5 to 2%)	Faridpur, Mymensingh, Comilla, Khulna, Dinajpur.
Group 3 (1 to 1.5%)	Jessore, Rajshahi, Rangpur, Barisal.
Group 4 (below 1%)	Sylhet.

Source: Based on Table 4.1.

Group 3 (1.0 to 1.5 per cent); Group 4 (below 1.0 per cent).

How closely are the districts ranked on the basis of production growth rates related to those ranked on the basis of yield growth rates? Are the growth rates in one period related to those in another? A casual observation of various groups of districts set out in Table 4.2 indicates that growth rates in yield and production within a phase are likely to be strongly related, while those between periods are unlikely to be so. As can be seen from Table 4.2, the districts of Kushtia, Khulna, Pabna, Dinajpur and Bogra, which ranked very low in respect of growth rates in Period 1, rank very high in Period 2. On the other hand, Rangpur, Sylhet and Chittagong Hill Tracts, which showed relatively high rates of growth in Period 1, have experienced very slow rates of growth in Period 2. For Chittagong Hill Tracts, the slow output growth is primarily due to a decline in cultivated area.

In order to test the strength of the relationships involving inter-temporal and intra-temporal growth rates in yield and production of foodgrains, Spearman rank correlation coefficients were calculated. As expected, intra-period growth rates in yield and production are strongly related. For instance, the coefficient of correlation between districts ranked on the basis of production and yield growth rates in Period 1 turned out to be 0.8547. The corresponding coefficient of rank correlation for Period 2 was 0.5735 which is appreciably lower than the one for the previous period. One needs to bear in mind that the growth rates in yield relate to those per gross cropped hectare. As argued elsewhere (Alauddin and Tisdell, 1986c), with the Green Revolution being more firmly established, output growth in food-grain may result more from increase in yield per net cropped culti-vated area, i.e. on an annual basis, than on yield per gross cropped area. This point is taken up for further analysis in the next section. As for inter-period rankings in respect of production or yield growth rates, there is little or no systematic relationship. Thus the rank correlation coefficient between growth rates in yields in Period 1 and those in Period 2 was –0.0147. For the inter-period production growth rates, the coefficient of rank correlation had a value of –0.0993.

4.3   INTER-DISTRICT GROWTH OF FOODGRAIN PRODUCTION AND YIELD IN THE PERIOD OF THE NEW TECHNOLOGY

This section is concerned with foodgrain production and yield in the period following the introduction of the Green Revolution tech-nologies. In estimating the growth rates of foodgrain yields, two types of crop yields have been distinguished: yield per gross cropped hectare (i.e. including multiple cropping) and yield per net cultivated hectare. It should be noted that district-level data on multiple crop-ping are available only for the 1969–84 period. This precludes any possibility of inter-temporal comparison of growth rates in yields per net cultivated hectare. Furthermore, as data on the latter type of yields were not readily available, they were derived using indices of multiple cropping to inflate appropriately the yields per gross cropped hectare.

Growth rates of foodgrain production and the two types of yields were estimated employing a semi-log function. The results for Bangladesh and its districts for the 1969–84 period are set out in Table 4.3. A comparison of the post-Green Revolution period in foodgrain production and (gross) yield set out in Table 4.1 with those in Table 4.3, indicates that the latter estimates are generally higher than those estimated by the kinked exponential model (3). However, the broad patterns of growth are very similar. The coefficients of rank correlation between the two estimates of production growth rates was 0.9118 while that for the two estimates of (gross) yields was 0.7595. The differences between the growth rates estimated for the same period reported in Table 4.1 and Table 4.3 may be attributed to discontinuity bias. This is because 'when growth trends for sub-periods of a time are estimated separately, there is no guarantee that the terminus of one trend line will coincide with the beginning of the next' (Boyce, 1986, p.390). In the present case, when the kinked exponential model is applied to the whole time series, the observations preceding the kink may not be entirely independent of those following them, as far as the estimation process is concerned. On the other hand, when a separate trend line is fitted to observations of the second sub-period, they are independent of the observations preceding the relevant sub-period.

One significant finding which emerges from an analysis of the information set out in Table 4.3 is that for most districts (as for Bangladesh as a whole) the growth rates in net yield exceed those in gross yield. In only four out of 17 districts are the growth rates of net yield lower than those in gross yield. Of these only in Chittagong Hill Tracts did growth rate in net yield significantly lag behind that in gross yield. To what extent are the growth rates in net yield related to those in production? The coefficient of correlation between ranks of districts on the basis of production growth rates and those on the basis of net yield growth rates was very high (0.8363). On the other hand, the corresponding coefficient between growth rates in gross yield and production had a much lower value (0.5620). This seems to confirm our findings in Chapter 2 that in the post-Green Revolution period foodgrain output grew primarily from an increase in yield on an annual basis.

Table 4.3 Growth rates of foodgrain production and yields per gross and net cropped hectare, Bangladesh districts, 1969–70 to 1984–5

District	Production				Yield/gross hectare				Yield/net hectare			
	Growth rate (%)	t-value	R^2	F-ratio	Growth rate (%)	t-value	R^2	F-ratio	Growth rate (%)	t-value	R^2	F-ratio
Dhaka	2.66[a]	8.07	0.823	65.14[a]	2.32[a]	8.37	0.833	70.07[a]	2.03[a]	6.62	0.758	43.83[a]
Mymensingh	3.57[a]	11.54	0.905	133.09[a]	2.13[a]	9.98	0.877	99.63[a]	2.79[a]	11.13	0.898	123.97[a]
Faridpur	2.31[a]	3.09	0.405	9.53[a]	2.11[a]	4.02	0.536	16.16[a]	1.83[a]	2.70	0.343	7.30[a]
Chittagong	2.61[a]	7.21	0.788	51.96[a]	1.81[a]	6.69	0.762	44.76[a]	3.13[a]	8.55	0.839	73.07[a]
Chittagong HT	0.69[d]	0.81	0.045	0.66[d]	2.72[a]	4.71	0.613	22.22[a]	0.64[d]	0.94	0.059	0.88[d]
Noakhali	2.54[a]	4.05	0.540	16.43[a]	2.74[a]	5.34	0.671	28.55[a]	3.22[a]	5.53	0.686	30.56[a]
Comilla	2.88[a]	3.77	0.503	14.18[a]	2.02[a]	4.82	0.624	23.21[a]	2.69[a]	4.61	0.602	21.22[a]
Sylhet	0.27[d]	0.28	0.006	0.08[d]	-0.09[d]	0.22	0.003	0.50[d]	0.82[c]	1.48	0.136	2.20[d]
Rajshahi	2.25[a]	3.72	0.497	13.82[a]	1.67[a]	3.85	0.515	14.86[a]	1.57[a]	3.10	0.407	9.62[a]
Dinajpur	3.51[a]	4.34	0.573	18.81[a]	1.82[a]	6.62	0.758	43.88[a]	2.06[a]	4.78	0.620	22.81[a]
Bogra	4.56[a]	9.73	0.871	94.63[a]	3.15[a]	8.63	0.842	74.51[a]	4.11[a]	9.43	0.864	88.85[a]
Rangpur	2.94[a]	7.14	0.784	50.93[a]	1.69[a]	4.56	0.597	20.76[a]	2.16[a]	5.28	0.666	27.92[a]
Pabna	4.23[a]	5.71	0.699	32.58[a]	2.77[a]	6.78	0.766	45.96[a]	3.22[a]	6.34	0.742	40.21[a]
Khulna	4.44[a]	4.76	0.617	22.61[a]	3.33[a]	4.88	0.629	23.80[a]	3.72[a]	4.61	0.603	21.23[a]
Kushtia	6.02[a]	6.72	0.763	45.12[a]	4.25[a]	7.34	0.794	53.86[a]	5.16[a]	7.18	0.787	51.61[a]
Jessore	2.87[a]	5.85	0.709	34.19[a]	2.32[a]	5.84	0.709	34.14[a]	2.91[a]	7.03	0.779	49.45[a]
Barisal	2.04[a]	3.76	0.503	14.16[a]	1.67[a]	3.09	0.405	9.55[a]	1.64[a]	3.00	0.392	9.01[a]
BANGLADESH	2.83[a]	7.70	0.809	59.37[a]	1.96[a]	8.42	0.835	70.90[a]	2.64[a]	7.32	0.793	53.63[a]

[a] Significant at 1 per cent level;
[b] Significant at 5 per cent level;
[c] Significant at 10 per cent level;
[d] Not significant.

Source: Based on district-level data from sources mentioned in Table 2.1 and BBS (1979, p.160; 1982, pp.166–7; 1984a, p.31; 1986a, p.39).

To sum up, the above analysis of the district-level growth in foodgrain production and yield clearly indicates significant variation in performance across districts within a particular phase. Performance of the same districts also varies between phases. While growth rates in production and those in yield within a phase are strongly related, there is little or no relation involving growth rates in either variable between sub-periods.

But the question is why do growth rates differ across districts? Are there any discernible differences in regional patterns in the diffusion of the different elements of the Green Revolution technology which could explain this? This is investigated in the next section.

4.4   FACTORS 'EXPLAINING' REGIONAL PATTERNS OF GROWTH IN FOODGRAIN PRODUCTION AND YIELD

This section investigates whether there are any inter-regional differences in the pattern of use of modern inputs which may have led to differential output performances by districts. This is because, as discussed in the preceding sections, growth in productivity per hectare has been the primary factor in the growth of foodgrain production in the post-Green Revolution period. In this respect, differences in inter-district diffusion of various elements of the new agricultural technology assume critical importance. Data on the input-use pattern, gross and net yields and other relevant variables are set out in Table 4.4. They relate to 1982–4, the average of the triennium ending 1984–5.

The information contained in Table 4.4 clearly indicates considerable inter-district differences in inputs used. For instance, fertilizer (kg of nutrients) applied per gross cropped hectare (FERT) ranges from about 14 kg for Barisal to 79 kg for Chittagong. Percentage of gross foodgrain area planted with HYVs (PRHYVF), percentage of foodgrain area irrigated (PRFAI), and percentage of gross cropped area irrigated by modern methods[2] (PCIMOD) show similar regional differences. *Kharif* (*aus* and *aman* rice) foodgrain area as a percentage of total *kharif* (rainfed) area (PCKHARIH) shows a much greater degree of variation across districts than the *rabi* (*boro* rice and wheat) foodgrain area as a percentage of total *rabi* area (PCRABIH). However, when *rabi* HYV are considered as a percentage of gross area planted with foodgrain (PCRABIHF), greater inter-district differences emerge. This does not happen with *kharif*

HYVs expressed as a percentage of gross foodgrain area (PCKHHF). If one considers the percentage of total rice area planted to HYVs of *boro* rice (PCBRHA), the inter-regional differences come into even sharper focus. Considerable regional variations seem to exist for foodgrain yields per net and gross cropped hectare (NYFDT and YFDT) as well as for *kharif* and *rabi* yields (KHARY and RABIY). The degrees of variation in NYFDT and KHARY are higher than those in YFDT and RABIY. This is probably because NYFDT is influenced by inter-district variation in cropping intensity (INTN). On the other hand, *boro* rice is entirely irrigated and wheat partly irrigated, compared to the primarily rainfed *kharif* foodgrain crops.

Variations in the use of these yield-augmenting inputs by districts are believed to have led to differences in the regional patterns of growth rates in production and yield of foodgrains. In order to assess the relative contribution of each of the yield-augmenting inputs on the two variants of foodgrain yields (YFDT and NYFDT) regression equations were estimated, the results of which are presented in Table 4.5. Regression equation 1 shows that PRFAI, PRHYVF and FERT have a very high explanatory power. However, the coefficient of PRFAI does not have the expected sign, while that of FERT is not significant at the 5 per cent level. A similar picture emerges if NYFDT is substituted as the dependent variable, as can be seen from equation 2. This is because of high collinearity between the explanatory variables.[3] It must be emphasized, however, that equation 2, which incorporates the same independent variables as equation 1, does have a much lower explanatory power. Subsequently cropping intensity (INTN) was introduced as an additional variable. After considering all possible subsets of regressions, the one involving FERT, PRHYVF and INTN, i.e. equation 3, appeared to be the best according to the BMDP P9R programme (see Dixon, 1983). A comparison of equation 2 with equation 3 clearly indicates the superiority of the latter equation in terms of both explanatory power and statistical significance of the coefficients. Because of high collinearity between FERT and PRHYVF, the coefficient of FERT is significant at the 10 per cent probability level. Thus multiple cropping has a significant positive impact on foodgrain yield on an annual basis (NYFDT).

Equations 4 to 9 in Table 4.5 show that each variable individually is highly significant, but only PRHYVF can explain a high percentage of variation in YFDT and NYFDT. Furthermore, it is the coefficient of PRHYVF which possesses highest *t*-value in either case. Thus

Table 4.4 Variables indicating district-level diffusion of new agricultural technology and agrarian structure in Bangladesh

District	FERT	PRHYVF	PRFAI	PCKHARIH	PCRABIH	PCIRMOD	PCKHHF	PCRABIHF	PCBRHA	SMALLOWN
Dhaka	68.37	33.07	20.25	14.85	86.69	16.08	11.09	22.00	18.70	0.439
Mymensingh	40.52	36.34	21.74	23.69	76.92	15.81	18.06	18.28	16.08	0.317
Faridpur	15.39	16.45	5.97	2.15	87.73	3.81	1.79	14.66	7.73	0.338
Chittagong	79.44	74.10	27.34	63.88	99.99	22.01	45.80	28.30	28.29	0.471
Chittagong HT	41.27	65.74	16.51	58.32	99.99	8.67	47.83	17.91	17.89	0.153
Noakhali	26.98	40.88	9.68	32.08	93.47	5.17	27.48	13.41	13.20	0.526
Comilla	68.37	43.24	17.20	26.71	94.03	13.26	20.15	23.08	15.75	0.541
Sylhet	18.77	17.66	21.19	10.91	34.12	8.86	7.72	9.94	9.35	0.267
Rajshahi	42.45	21.75	15.43	8.24	92.10	10.34	6.91	14.84	7.46	0.224
Dinajpur	44.04	23.79	8.57	13.10	98.99	7.54	11.38	12.41	2.23	0.160
Bogra	74.77	39.09	20.08	23.35	99.54	18.01	18.51	20.57	15.14	0.317
Rangpur	28.26	24.13	11.04	13.35	95.16	8.22	11.58	12.55	5.16	0.266
Pabna	43.14	24.89	13.98	5.19	92.78	10.36	4.03	20.86	11.23	0.264
Khulna	18.51	15.11	6.75	11.53	71.04	5.49	10.85	4.26	3.80	0.276
Kushtia	65.94	41.45	35.53	25.16	97.02	26.58	19.37	22.08	1.63	0.232
Jessore	38.26	23.85	14.83	12.67	94.29	11.53	10.90	12.95	7.45	0.238
Barisal	13.85	12.86	7.47	8.96	78.91	4.27	8.47	4.39	4.12	0.383
CV (%)	50.39	52.66	48.90	82.74	18.49	55.47	78.87	40.70	65.41	36.873

District	INTN	YFDT	NYFDT	KHARY	RABIY	GINIOWN	GINIOP	POTKHYV	SHARET	SMALLOP
Dhaka	141.9	1545	2192	1174	2637	0.472	0.519	27.08	14.55	0.412
Mymensingh	176.4	1505	2656	1198	2486	0.491	0.517	53.17	15.55	0.304
Faridpur	159.3	1070	1703	758	2608	0.471	0.526	13.10	16.88	0.297

Chittagong	159.6	1996	3185	1781	2550	0.445	0.512	99.18	30.06	0.435
Chittagong HT	128.3	1740	2231	1561	2573	0.380	0.400	84.04	25.64	0.141
Noakhali	154.7	1472	2277	1290	2555	0.451	0.554	69.81	30.40	0.441
Comilla	168.3	1590	2676	1297	2488	0.456	0.485	34.83	9.93	0.531
Sylhet	143.0	1322	1888	1141	1762	0.512	0.556	66.85	14.79	0.236
Rajshahi	128.8	1335	1721	1138	2367	0.490	0.509	57.96	20.53	0.207
Dinajpur	137.7	1417	1952	1313	2144	0.477	0.456	98.20	19.70	0.171
Bogra	173.1	1684	2918	1425	2682	0.501	0.504	72.72	15.02	0.323
Rangpur	187.3	1431	2680	1301	2298	0.514	0.504	71.07	14.77	0.280
Pabna	160.0	1239	1983	853	2595	0.491	0.523	28.71	16.21	0.242
Khulna	127.1	1386	1762	1339	2142	0.498	0.559	54.85	23.30	0.223
Kushtia	142.7	1428	2039	1122	2506	0.496	0.530	91.18	14.15	0.204
Jessore	135.9	1289	1746	1092	2508	0.454	0.491	49.95	14.16	0.208
Barisal	139.5	1216	1695	1160	2153	0.451	0.570	62.57	32.40	0.290
CV (%)	12.09	15.16	21.37	19.31	10.03	6.91	8.01	41.13	34.77	36.996

Notes: FERT: chemical fertilizers (kg of nutrient/ha of gross cropped area); PRHYVF and PRFAI: percentage of foodgrain area under HYVs and irrigation; PCKHARIH and PCRABIH: percentages of HYV area in *kharif* and *rabi* foodgrains; PCIMOD: percentage of gross cropped area irrigated by modern methods (tubewells, low lift pumps and large-scale canals); PCKHHF, PCRABIHF and PCBRHA: percentages of HYV area of *kharif*, *rabi* and *boro* rice area in total foodgrain area; INTN: intensity of cropping; YFDT and NYFDT: yield per gross and net cropped hectare in foodgrains; KHARY and RABIY: *kharif* and *rabi* foodgrain yields; GINIOWN and GINIOP: Gini concentration ratios for owned and operated holdings in 1983–4; POTKHYV: percentage of *kharif* foodgrain area potentially suitable for HYV cultivation; SHARET: percentage of cultivated area under sharecropping in 1976–7; SMALLOWN and SMALLOP: proportions of owned and operated area by small farms (< 1 ha) in 1983–4 in relation to total area. All figures unless otherwise specified relate to 1982–4 (average for the years 1982–3, 1983–4 and 1984–5); CV is coefficient of variation.

Sources: Based on district-level data from sources mentioned in Table 2.1; BBS (1979, pp.162, 166–7, 212); BBS (1981, pp.61, 71, 81, 91, 101, 121, 131, 141, 151, 161, 181, 191, 201, 211, 221, 241, 251, 261, 271); BBS (1982, pp.206, 209, 213); BBS (1984a, pp.31, 33); BBS (1986a, pp.39, 42–3, 70, 72); BBS (1986b, pp.141–61); Bramer (1974, pp.65–6).

Table 4.5   Linear regression analysis of determinants of foodgrain yields per gross and net area cropped with foodgrains (YFDT and NYFDT), using Bangladeshi district-level cross-sectional data, 1982–4

Dependent variable	Intercept	FERT	PRFAI	Independent variable PRHYVF	INTN	R^2	F-ratio	DF
YFDT	1057.53	2.9909 (1.63)[c]	−3.4065 (0.75)[d]	9.8131 (4.94)[a]	–	0.8334	21.68[a]	3,13
NYFDT	1487.25	8.7750 (1.40)[c]	−8.1447 (0.53)[d]	14.1728 (2.09)[b]	–	0.5745	5.85[a]	3,13
NYFDT	−709.03	4.2236 (1.58)[c]	–	13.5776 (4.09)[a]	15.1169 (6.37)[a]	0.8945	36.74[a]	3,13
YFDT	1135.86	7.3529 (4.04)[a]	–	–	–	0.5206	16.29[a]	1,15
YFDT	1217.51	–	14.5024 (2.35)[b]	–	–	0.2691	5.52[b]	1,15
YFDT	1077.46	–	–	11.4501 (7.72)[a]	–	0.7991	59.66[a]	1,15
NYFDT	1583.94	14.2477 (3.37)[a]	–	–	–	0.4302	11.32[a]	1,15
NYFDT	1770.75	–	26.3239 (1.91)[b]	–	–	0.1951	3.64[c]	1,15
NYFDT	1559.84	–	–	19.4558 (3.93)[a]	–	0.5078	15.47[a]	1,15

[a] Significant at 1 per cent level;
[b] Significant at 5 per cent level;
[c] Significant at 10 per cent level;
[d] Not significant.
DF is degrees of freedom.

Source: Based on Table 4.4

while both irrigation and chemical fertilizers have significant positive impact on crop yields, *the HYV area variable (PRHYVF) emerges as the most dominant determinant*. This is further substantiated from the values of the rank correlation coefficients of YFDT with PRHYVF, FERT and PRFAI respectively are 0.8358, 0.6806 and 0.5784. The values of the coefficients of rank correlation of NYFDT with these variables provide similar indications.

4.4.1   Factors Affecting Expansion of Area Under HYV

The area under HYV is critical for foodgrain yield levels. What factors determine the expansion of area under HYV cultivation? Both technological and agrarian structural variables are generally believed to influence the growth process in Bangladeshi agriculture:

Technological factors: (1) area under irrigation, (2) proportion of land suitable for HYV cultivation, and (3) fertilizer.

Agrarian structural factors: two elements in the structure appear to be: (1) land ownership and distribution pattern, and (2) land tenure arrangements.

One would expect the technological factors to influence positively the expansion of HYV area, while the influence of the agrarian structural factors on the dependent variable is difficult to predict *a priori*. Some of the arguments have been presented by Alauddin and Mujeri (1986a) and Mahabub Hossain (1980). We do not wish to repeat them here.[4]

We now use the information in Table 4.4 to assess the influence of these two sets of variables on the spread of HYVs. Apart from identifying determinants of the aggregate extent of HYV adoption (PRHYVF), those 'explaining' district-level variation in HYV adoption in *rabi* and *kharif* seasons (PCRABIFH and PCKHHF), are also isolated. Our regression estimates are set out in Table 4.6. The estimates reported here are only those that are considered the 'best' from all possible subsets of regressions (Dixon, 1983).

Let us now consider the regression results presented in Table 4.6. As expected, the percentage of gross cropped area irrigated by modern methods (PCIMOD) is the most important determinant of the intensity of *rabi* HYV expansion (PCRABIHF). The percentage area irrigated in foodgrains (PRFAI) has a significantly positive impact on overall expansion of the HYV area (PRHYVF) as well as the *kharif* HYV area (PCKHHF). PRHYVF and PCKHHF are also significantly determined by the percentage of *kharif* foodgrain area potentially suitable for HYV cultivation (POTKHYV). Along with the significantly positive impact of PRFAI, PCKHHF is also seen to be positively influenced by the percentage of *boro* HYV area as a proportion of total foodgrain area (PRBHA). This may have two implications. First, in drought-prone districts primarily those farmers that have access to irrigation adopt HYVs during the *kharif* season, so that in case of inadequate and uncertain rainfall supplementary irrigation can be arranged. This is supported by farm-level evidence (Alauddin and Tisdell, 1988e; see also Chapter 7). Secondly, farmers' experience with *boro* HYV cultivation is likely to have a positive impact on the decision to adopt HYVs during the *kharif* season.

The variables used as proxies for agrarian 'social' structure also

Table 4.6 Linear regression analysis of determinants of HYVs of all, *rabi* and *kharif* foodgrains (PRHYVF, PCRABIHF and PCKHHF), using Bangladeshi district-level cross-sectional data, 1982–4

Explanatory variable	Dependent variable											
	PCRABIHF		PCRABIHF		PCKHHF		PCKHHF		PRHYVF		PRHYVF	
	coeff.	t-value	coeff.	t-value	coeff.	t-value	coeff.	t-value	coeff.	t-value	coeff.	t-value
PCIMOD	0.7188	4.66[a]	0.6631	5.48[a]	–	–	0.8470	4.67[a]	–	–	–	–
PRFAI	–	–	–	–	–	–	–	–	1.1092	3.58[a]	1.3436	7.42[a]
GINIOWN	–74.2732	2.26[b]	–	–	–158.3820	4.12[a]	–	–	–292.7470	4.15[a]	–	–
GINIOP	–	–	–64.5531	3.37[a]	–	–	–201.5990	5.78[a]	–	–	–271.3910	7.80[a]
SMALLOWN	13.1321	1.61[c]	–	–	–	–	37.29	2.82[a]	35.1231	1.73[c]	–	–
SMALLOP	–	–	25.2721	3.41[a]	–	–	–	–	–	–	66.8692	5.07[a]
SHARET	–0.2823	1.67[c]	–	–	–	–	1.32	6.21[a]	–	–	1.2784	6.00[a]
PRBHA	–	–	–	–	0.9036	5.17[a]	–	–	–	–	–	–
POTKHYV	–	–	–	–	0.2746	6.07[a]	–	–	0.1866	1.82[b]	–	–
R^2	0.7566		0.8281		0.9046		0.8698		0.7991		0.9249	
F-ratio	9.320[a]		20.870[a]		41.080[a]		20.050[a]		11.930[a]		36.930[a]	
Degrees of freedom	4,12		3,13		3,13		4,12		4,12		4,12	

[a] Significant at 1 per cent level;
[b] Significant at 5 per cent level;
[c] Significant at 10 per cent level.

Source: Based on Table 4.4.

significantly affect expansion of HYV area. For instance, districts with higher Gini concentration ratios for owned or operated land holdings (GINIOWN or GINIOP) are likely to have lower percentages of foodgrain area under HYVs either seasonally or annually. On the other hand, districts with a higher percentage of land owned or operated by small farmers (under 1 hectare, SMALLOWN and SMALLOP) are likely to have a higher percentage of foodgrain area under HYVs. However, SMALLOP is a more significantly positive determinant of HYV area expansion than SMALLOWN, as can be seen from the statistical significance of the coefficients in the relevant equations. Thus a greater degree of inequality in the distribution of land ownership or control seems to be an impediment to widespread adoption of agrarian innovations. Our findings seem to be consistent with those of Mahabub Hossain (1980) in this regard. This is also supported by farm-level evidence (see, for example, Barker and Herdt, 1978; Hayami and Ruttan, 1984; Parthasarathy and Prasad, 1978; for further details see Chapter 6).

The variables used as proxies for mode of production in Bangladesh, namely, the percentage of cultivable land under sharecropping (SHARET) has a weak negative impact on PCRABIHF but has significant positive impact on expansion of overall *kharif* HYVs (PRHYVF and PCKHHF). This may be explained as follows: SHARET and PCIMOD are negatively related as indicated by the correlation (−0.303). On the other hand, PCIMOD is the most important determinant of PCRABIHF. Furthermore, the cultivation of *boro* HYV and other crops under sharecropping arrangements appears to be less rewarding for sharecroppers because of high costs of inputs including irrigation water. The cultivation of HYV *aus* and *aman* rice, on the other hand, mostly takes place under rainfed conditions where farmers have significant cost advantages relative to *boro* HYV rice and HYV wheat (Alauddin and Mujeri, 1986a, pp. 68–9; see also Chapter 6). Furthermore, on a full-cost basis per hectare, net returns to sharecroppers from all cereal crops except broadcast *aus* are negative, while on a cash-cost basis net returns from *aus* and *aman* HYVs are positive and those from the *rabi* cereals are negative.[5] Alauddin and Mujeri (1986a, p.69) also reported that net returns per hectare were higher for *aus* and *aman* HYVs compared to those of the local varieties of these crops.

M. Hossain (1980) shows unfavourable impact of tenancy on the adoption of HYVs. This may have resulted from the fact that his tenancy data relate to 1960 while his HYV area data relate to

1977–8. It is not realistic to use the 1960 data for one variable to explain another almost two decades later. Changes in the agricultural sector at the advent of the new technology and demographic and subsistence pressure appear to have changed the development perspective over the years (for details see Alauddin and Mujeri, 1986a, pp.68–9).

4.5 ARE REGIONAL GROWTH RATES OF FOODGRAIN PRODUCTION AND YIELD MORE DIVERGENT IN THE POST-GREEN REVOLUTION PERIOD?

Are regional growth rates of foodgrain yield and production more divergent in the post-Green Revolution period than in the period preceding it? When the coefficients of variation of growth rates of foodgrain yields and production in Period 1 are compared with those in Period 2, it can be observed that the coefficient of variation of growth rates of foodgrain production fell by more than 33 per cent (from 65.04 per cent in 1947–68 to 43.54 per cent in 1969–84). At the same time, the coefficient of variation of growth rates of foodgrain yield declined by nearly 53 per cent (from 81.14 in 1947–68 to 38.21 per cent in 1969–84). This means that the *relative* degree of variation in growth rates between districts has fallen.

It would also be useful to consider whether the growth rates are absolutely less divergent as well as relatively so. To investigate this possibility one can employ standard deviations of the growth rates for the two periods. It was found that the standard deviation of production growth rates has declined by 6.56 per cent (from 1.22 per cent in 1947–68 to 1.14 per cent in 1969–84), while that of yield growth rates fell by 11.39 per cent (from 0.79 to 0.70 per cent) over the same period. This clearly indicates that the growth rates in foodgrain production and yield at the district level are less divergent in Period 2 than in Period 1. This seems to suggest that on the whole, *the Green Revolution has had a moderating impact on the inter-regional variation in growth rates*.

To what extent does this evidence support the hypothesis that the Green Revolution has reduced regional imbalances in foodgrain production in Bangladesh? Consider the coefficients of variation of district-level average yields in Period 1 and Period 2. The relative measures of dispersion of (gross) yields recorded an increase of nearly 88 per cent between periods (from 8.72 per cent in 1947–68 to

16.37 per cent in 1969–84). In order to obtain some further evidence, coefficients of variation of gross and net yields (YFDT and NYFDT) of 1969–70 were compared with those of 1982–4 (average for the years 1982–3, 1983–4 and 1984–5). The coefficient of variation of YFDT declined by more than 17 per cent (from 18.38 per cent in 1969–70 to 15.16 in 1982–4) while that of NYFDT remained unchanged at around 21 per cent.

But why this apparently conflicting evidence? One plausible explanation could be slow growth rates in yields in the pre-Green Revolution period compared to the period following it. This is reflected in the values of the ranges (differences between the highest and the lowest values). This range increased dramatically from 314 kg/hectare in 1947–68 to 950 kg/hectare in 1969–84. On the other hand, it was 850 kg/hectare in 1969–70 which was only marginally lower than that in 1982–4 (926 kg/hectare). However, faster growth rates in yields in districts with lower base values (e.g. Kushtia, Pabna, Dinajpur, Khulna) and relatively slower growth rates in districts with higher base values (e.g. Chittagong) seemed to reduce the gap somewhat so that by the end of Period 2, the individual yield values appeared relatively closer to their combined average compared to those in 1969–70. Thus it seems possible that, with the passage of time, the inter-regional differences in growth of foodgrain production have narrowed to some extent (see also Chapters 9 and 10).

Let us now compare our results with those employing district-level growth rate data from the Indian state of Andhra Pradesh (Parthasarathy, 1984). An analysis of the district-level foodgrain production growth rates in the pre- and post-Green Revolution periods (Parthasarathy, 1984, p.A75) suggests that the standard deviation of growth rates fell by 20.65 per cent (from 2.329 per cent to 1.848 per cent). The coefficient of variation registered even a greater decline – 53.47 per cent over the same period. Thus the evidence from Andhra Pradesh is consistent with the findings of the present study.

4.6 CONCLUDING COMMENTS

Overall growth rates in production and yield of foodgrains are higher in the post-Green Revolution period compared to the period preceding it. However, the picture of overall crop output growth for Bangladesh as a whole conceals important inter-district differences in growth rates of production and yield. Intra-period growth rates in these two variables are strongly correlated. However, there is little or no correlation between them when inter-temporal comparisons are made. There is evidence that divergence in inter-district growth rates has been moderated to some extent in the post-Green Revolution period.

Considerable differences in the pattern of modern input use have led to regional variations in output and productivity growth. Significant regional variations in the spread of the new technology seem to have resulted in similar variations in intensity of cropping. While both irrigation and chemical fertilizers have significant positive impact on the increase in foodgrain yields, it is the percentage area under HYV which emerges as the most dominant determinant of the increase in overall output per hectare.

Turning to the factors that have contributed to the expansion of HYV area, it is found that irrigation and potentially suitable HYV *aus* and *aman* area make a significant positive impact. The socio-economic variables appear to have considerable influence on the intensity of HYV adoption. The findings point to two aspects of the growth process. First, land ownership/control concentration is a serious obstacle to expansion of HYV area and hence impedes the pace of growth. Secondly, under the present system of tenural arrangements, income inequality is likely to be further accentuated because of differential gains between the owner and the share-cropper.

5 Reduced Diversity of Crops, Narrowing of Diets and Changes in Consumer Welfare

5.1 INTRODUCTION

Growth in economic production does not necessarily imply an increase in economic welfare. Although, as shown in the last three chapters, foodgrain production has risen substantially in Bangladesh as result of the Green Revolution this appears to be have been at the expense of non-cereal crops (see Chapter 2). Despite the rapid growth in Bangladeshi foodgrain production, it is possible that Bangladeshi food supply has become less diversified and that economic welfare has declined in some way. Also protein deficiency may have increased owing to the slow growth rate in production of pulses. The quality of the average Bangladeshi diet may have fallen. If so this could be a possible cost of the Green Revolution. Redclift (1987, p.64) mentions the narrowing of popular diets as one of the possible costs of the Green Revolution. Let us explore this matter for Bangladesh.

5.2 TRENDS IN PER CAPITA AVAILABILITY OF CEREAL AND NON-CEREAL COMMODITIES IN BANGLADESH, 1967–8 TO 1984–5

First let us consider trends in the per capita availability of cereal and non-cereal food commodities in Bangladesh since the introduction of the Green Revolution technologies. Table 5.1 sets out data on the per capita availability of cereal and non-cereal food commodities. It can be seen that overall per capita availability of cereals has not increased during the last two decades, while per capita availability of pulses, fruits and spices has shown a marked decline. With regard to a balanced diet, the position in relation to pulses, fruits and vegetables is particularly worrying.

91

Table 5.1   Per capita availability of cereal and non-cereal food
commodities: Bangladesh, 1967–8 to 1984–5

| Year | Cereals | Kg per head of population | | | |
		Pulses	Vegetables	Fruits	Spices
1967	168.547	3.407	23.904	23.409	4.916
1968	172.055	3.606	25.080	21.691	4.646
1969	177.238	3.618	25.849	20.094	4.551
1970	161.696	3.612	24.918	17.639	4.368
1971	148.953	3.385	21.409	16.257	3.856
1972	159.077	2.608	20.078	15.089	3.559
1973	163.618	2.388	18.514	14.893	3.244
1974	152.819	2.515	21.120	14.330	3.238
1975	162.980	2.414	21.821	14.084	3.239
1976	146.206	2.485	18.878	14.014	2.899
1977	168.303	2.505	20.334	13.559	3.036
1978	165.284	2.336	20.629	12.692	2.854
1979	163.947	2.166	20.209	12.596	2.835
1980	155.541	2.069	19.729	12.812	2.317
1981	162.717	1.951	21.259	12.769	2.575
1982	165.858	2.016	20.731	10.781	2.647
1983	166.285	1.842	20.545	11.720	2.533
1984	169.135	1.760	19.519	11.550	2.484

Notes: 1967 refers to 1967–8 (July 1967 to June 1968), etc.

Source: Based on data from sources mentioned in Table 2.1 and Alauddin
and Tisdell (1988b, Table 3.1).

Figure 5.1 highlights the changing situation in relation to the per
capita supply of the main food commodities following the Green
Revolution. Availability of each of the components is shown relative
to its supply in 1967–8 which is taken as the base year and assigned an
index of 100. A massive decline in per capita availability of pulses,
fruits and spices is evident, as well as a significant decline in the
availability of vegetables. Per capita production of pulses in 1984–5 is
only 51.7 per cent of that in 1967–8, and respectively for vegetables,
fruits and spices in 1984–5 only 81.7 per cent, 49.4 per cent and 50.5
per cent of that in 1967–8. On the other hand, per capita availability
of cereals in 1984–5 was hardly any more than in 1967–8, only 100.4
per cent of the 1967–8 figure. In fact the 1984–5 cereal supply figure is
unrepresentative because on the whole there has been some tendency
for per capita production of cereals to fall slightly in Bangladesh
(Alauddin and Tisdell, 1988b; see also Chapter 12).

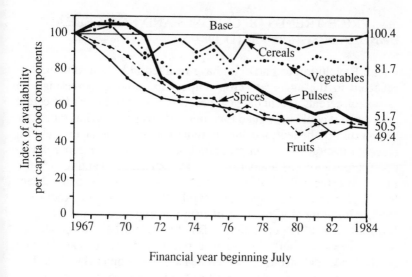

Financial year beginning July

Figure 5.1   Trends in per capita availability of main food crops in
post-Green Revolution Bangladesh, 1967–8 to 1984–5

Bangladeshi experience with changes in the composition of farm
production following the new technology is similar to that of India.
According to Staub and Blase (1974, pp.584–5).

between 1964–65 and 1970–71. . .cereal production rose by 26.5
per cent while the production of pulses and fibres declined and
oilseed production increased slightly. The increased production of
cereals relative to other crops has been due to (1) higher yields for
rice and wheat, (2) a substitution of high-yield cereals for other
crops, and (3) and an expansion of gross area cultivated.

More recent evidence, (Sawant, 1983) indicates that this bias still
persists. For instance, between 1968–9 and 1980–1, Indian rice and
wheat production grew at compound annual rates of 2.12 and 6.03
per cent respectively. At the same time gram and *tur*, two important
types of pulses, recorded compound annual growth rates of −0.89
and 0.60 per cent respectively. Growth rates in yields per hectare of
these crops showed similar unevenness (Sawant, 1983, pp.479, 491).

5.3   PRICE TRENDS OF CEREALS AND NON-CEREALS, BANGLADESH, 1969–70 TO 1984–5

Green Revolution technologies have resulted in 'commodity-bias' as reflected in the aggregate crop output-mix moving increasingly in favour of cereals. Technological progress has been fastest for cereal crops, and it could be argued that much of this progress has indirectly been a spin-off benefit, albeit sometimes via international agricultural research institutions, from progress in cereal culture in more developed countries in the temperate regions such as Japan and the United States. But technological progress has been slow or virtually absent for most non-cereal food-crops (Gill, 1983; Pray and Anderson, 1985). Consequently, because of rising aggregate demand owing to Bangladesh's rising population (BBS, 1985b, p.636) non-cereal food prices have risen both relative to cereal prices and in real absolute terms. Since real incomes on average have not risen in Bangladesh in recent years, this means that most of the population, particularly the poor, have been forced to rely increasingly on cereals to fill the bulk of their diet. Cereals now account for a larger proportion of the popular diet and non-cereals a smaller proportion.

The elasticity of demand with respect both to income and prices for cereals is much lower than for non-cereal food items. We do not have separate price elasticities for cereal and non-cereal food items. Bangladesh being very much a subsistence economy and cereals (primarily rice) being the staple food, one would *a priori* expect a lower degree of sensitivity to changes in their prices (Alauddin and Tisdell, 1986d) compared to non-cereal food items, e.g. pulses. The 1981–2 household expenditure survey (BBS, 1986e, p.38) estimates the income elasticities of demand for foodgrains to be much lower than those for non-cereal food items. For instance, rice has an income elasticity of demand of 0.53 as against that of 1.48 for pulses for rural areas.

Price trends can be substantiated from the information set out in Table 5.2. The columns marked with IPs as prefixes refer to retail price indices, i.e., retail price indices deflated by wholesale food price index (FOODCPI), while those prefixed with RPs are real retail price indices relative to that of rice (IPRICE). LENTL, POT, CHIL, ONI, MUST, CHICK and BAN respectively refer to lentils, potatoes, chillies, onions, mustard oil, chicken and bananas. These are used as proxies for various non-cereal food items.

There seems to have been a slight downward trend in rice prices in real terms; non-cereal food items, while differing widely among themselves, have, except possibly for potatoes, registered significant increases in prices in recent years. It can also be observed that real price movements of non-cereals have shown a greater degree of instability (variability) than cereals. For instance, the coefficients of variation of real retail prices of lentils (IPLENTL), fish (IPFISH) and beef (IPBEEF) are respectively 18.01, 20.00 and 15.18 per cent compared to 8.59 per cent for the real price of rice (IPRICE). A similar picture emerges if one compares the relative price indices.

While the real price of cereals has fallen, this fall has been small and has not had a significant income effect. This, combined with the very considerable increase in the real and relative price of non-cereal food items, has meant that the forces of substitution of commodities in consumption have been working in favour of cereals. In addition, most Bangladeshis are forced to consume all their income because they are near subsistence level and practically all of this income is spent in purchasing food. Their income constraint plus the pattern of price changes have forced them to reduce their intake of non-cereal food items. There has not been sufficient scope for instance for the 'Giffen effect' to dominate. This can be illustrated in terms of the Hicksian indifference curve technique.

Two alternative scenarios can be considered depending upon how one interprets the real retail price index of rice (IPRICE) which is used as a proxy for the price index of cereal food items taken as a whole. One interpretation is that it shows a slight declining tendency. If on the other hand the quality of data is taken into account and some caution is exercised in interpreting the data, another possible interpretation is that real rice prices have remained *more or less* constant.

Against this background, indifference curves and budget lines are drawn as shown in Figure 5.2 assuming the latter interpretation. Figure 5.2 shows the variation in the consumption pattern of a *typical* (hypothetical) Bangladeshi consumer following the Green Revolution. The consumer's equilibrium shifts from position A to B, given that the consumer's budget line shifts from ML to $M'L$. Note that welfare has fallen, since the consumer is on a lower indifference curve at B. Figure 5.2 is drawn on the assumption that the real price of cereals has remained more or less constant and non-cereals have risen in prices, as is the case and can be seen from Table 5.2.

From a welfare point of view, however, it makes little difference how the budget line of the typical consumer has shifted. In the pre-

Table 5.2 Food price index, real and relative price indices of cereal and non-cereal food commodities: Bangladesh, 1969–70 to 1984–5 (with 1969–70 = 100)

Year	FOODCPI	IPRICE	IPLENTL	IPPOT	IPCHIL	IPONI	IPMUST	IPFISH	IPBEEF	IPCHICK	IPBAN
1969	100	100	100	100	100	100	100	100	100	100	100
1972	187	105	120	95	53	68	140	111	116	117	114
1973	257	103	149	113	106	123	130	110	118	105	110
1974	508	106	88	75	242	69	137	67	81	66	72
1975	343	99	163	115	92	109	137	117	126	117	115
1976	323	91	131	88	134	67	124	144	133	136	112
1977	377	99	153	88	203	184	144	137	133	134	110
1978	430	95	155	80	103	76	126	153	154	143	109
1979	528	104	147	76	60	127	101	142	131	122	113
1980	545	87	180	89	149	106	127	148	142	133	120
1981	624	75	156	77	127	201	105	140	141	126	111
1982	633	101	166	59	68	135	102	148	135	129	105
1983	725	95	151	75	126	101	121	155	120	138	104
1984	817	97	129	67	137	108	113	158	146	155	114

Year	IPMILK	RPLENTL	RPPOT	RPCHIL	RPONI	RPMUST	RPFISH	RPBEEF	RPCHICK	BPBAN	RPMILK
1969	100	100	100	100	100	100	100	100	100	100	100
1972	86	115	90	51	65	134	105	110	111	109	82

Year											
1973	88	144	110	102	119	126	107	114	102	106	85
1974	75	83	70	228	65	129	63	76	62	68	71
1975	136	165	117	93	110	139	118	127	119	116	137
1976	147	144	97	147	74	136	158	146	149	123	161
1977	128	154	89	205	186	145	139	135	135	111	129
1978	120	164	85	109	80	133	161	163	151	116	127
1979	108	141	73	58	122	97	137	127	117	109	104
1980	118	207	103	172	123	147	171	164	153	139	136
1981	121	208	102	169	268	140	188	188	168	147	161
1982	119	165	58	67	134	101	147	134	127	104	117
1983	109	159	78	133	106	127	163	126	145	109	114
1984	116	133	69	142	112	117	163	151	160	118	121

Notes: 1969 means 1969–70 (July 1969 to June 1970), etc.; FOODCPI refers to wholesale food price index. Columns with IPs as prefixes refer to real retail price indices, i.e. retail price indices deflated by FOODCPI. Columns prefixed with RPs are real retail price indices deflated by real retail price index of rice (IPRICE). LENTL, POT, CHIL, ONI, MUST, CHICK, BAN respectively refer to lentils, potatoes, chillies, onions, mustard oil, chicken and bananas. The remaining acronyms are self-explanatory.

Source: Based on data from BBS (1979 p.370, 372; 1985b, p.38, 640).

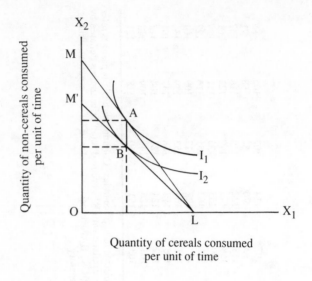

Figure 5.2   Following the Green Revolution, the typical (hypothetical) Bangladeshi consumer appears to have moved from an equilibrium represented by point *A* to one indicated by point *B* and has suffered a decline in economic welfare

Green Revolution phase, per capita consumption of all the major food crop components except cereals was higher in Bangladesh, with cereals being approximately the same as before the Green Revolution. Hence in equilibrium and according to the usual axioms of consumer theory (Samuelson, 1948; Tisdell, 1972) the typical Bangladeshi consumer is now worse off than prior to the Green Revolution.

This suggests that while more people are now able to survive in Bangladesh as a result of the Green Revolution, their diet is in general more restricted and less nutritious. The 1981–2 nutrition survey of Bangladesh (Ahmad and Hassan, 1983, p.31) indicates that per capita daily consumption of animal products and other plants and vegetables as sources of protein significantly declined (from about 18 g to about 11 g between 1962–4 (pre-Green Revolution) and 1981–2 (post-Green Revolution). Ahmad and Hassan (p.22) provide evidence also of a progressive decline in per capita calorie intake over the years. For instance, calorie intake in 1981–2 was 1943, a decline of 7 per cent from 2094 in 1975–6 and by 16 per cent from 2301 in 1962–4. On the face of it, therefore, Bangladeshi welfare per head is lower than in the past.

In Bangladesh, all the potential gains from the Green Revolution appear to have been swallowed up in supporting a larger population rather than in improving the standard of living, which appears to have declined for the masses. The position is even more worrying than it appears at first sight because legitimate doubts have been raised about whether Bangladesh's cereal production can be sustained because of its increasing dependence on the use of non-renewable (and to a considerable extent imported) resources and because of adverse environmental trends stemming from intensification of land use and chemicals which may adversely affect production in the future (Alauddin and Tisdell, 1987; 1988b; see also Chapter 13).

5.4 CONCLUDING COMMENTS

Although crop production overall has increased markedly in absolute terms since the onset of the Green Revolution in Bangladesh, this increase has been due mainly to expanded production from food crops, especially of cereals, rice and wheat. Increased incidence of multiple cropping of cereals in monoculture appears to have strongly contributed to this result.

Despite the substantial increase in food production which has enabled many more mouths to be fed in Bangladesh, there is cause for concern. Monocultural multiple cropping of cereals may prove to be ecologically unsustainable or less sustainable than in the past and there are already signs that it is becoming more costly and difficult to maintain productivity using such cultural practices (Hamid *et al.*, 1978, p.40; Alauddin and Tisdell, 1987; 1988b; Tisdell and Alauddin, 1989).

Growth in the level of Bangladeshi cereal production has not been able to match population increases and Bangladesh has become more dependent on food imports (Alauddin and Tisdell, 1988b). Not only has food consumption per head in Bangladesh failed to increase (it seems to have declined somewhat), it has also become less varied. The average Bangladeshi has increased his/her dietary dependence on rice and cereals and has been forced to reduce consumption of pulses, fruits and vegetables and protein-rich foods. Thus impoverishment in Bangladesh has had a qualitative dimension. Whether this will be a passing phase and there will be a greater variety of food of higher protein content in the future remains to be seen. However, the prospects for this seem dismal unless by some miracle Bangladesh can extricate itself from its present low-level equilibrium growth trap.

6 Market Analysis, Surpluses, Income Distribution and Technologically Induced Shifts in Supply

6.1 INTRODUCTION

The discussion in the previous chapter led to the conclusion that the welfare of the average consumer in Bangladesh has not increased since the Green Revolution and indeed in all probability has declined somewhat. Nevertheless, the Green Revolution has enabled more individuals to survive in Bangladesh, and if one places a positive value on the survival of a greater number of individuals then welfare has increased overall.

On the other hand, if the model of Hayami and Herdt (1977) is applied to the situation, one reaches the conclusion that the impact of the Green Revolution has been quite positive in increasing the welfare of both consumers and producers, if rice production or foodgrain production is made the basis of the modelling. While we shall apply the Hayami-Herdt model to the Bangladeshi situation, we have doubts about its value except in a limited context. It may indicate welfare gains which are illusory and mask important changes in income distribution which have occurred in Bangladesh with the Green Revolution.

The model employed by Hayami and Herdt, the Hayami-Herdt (H-H) approach, and subsequently used by Hayami and Ruttan (1985) is adopted primarily to test the adequacy of this type of model for analysing distributional consequences of technological change (Alauddin and Tisdell, 1986d). Furthermore, this model has been applied to the Philippines rice economy and it is useful to make a comparison between the Philippines and Bangladesh. The present chapter critically reviews the H-H model. A number of shortcomings are identified, and it is contended that inappropriate economic infer-

100

ences can be drawn from it. After the basic model is presented in terms of geometry and algebra, it is applied to Bangladeshi data. Shortcomings are then considered and their significant limitations are illustrated by Bangladeshi data. It is argued that the H-H model fails to provide a realistic assessment of the income distributional consequences of the Green Revolution.

6.2   ANALYTICAL FRAMEWORK: THE H-H MODEL

The investigation into the impact of technological change on the distribution of income within the framework of a market analysis model involves an examination of the distribution of welfare gains between (1) consumers and producers, (2) small and large producers and (3) rural and urban consumers. To provide an analytical background to the Bangladeshi context the H-H model is briefly outlined (for details see Hayami and Herdt, 1977, pp.246–9).

6.2.1   Consumers vs Producers

Figure 6.1A presents market demand and supply curves as well as the demand curve of the producers for home consumption of a semi-subsistence agricultural commodity $(D_H H)$. $D_H D_M D$ represents the total demand curve with $D_M D$ as the market demand curve. The quantity purchased by the non-farm households is measured by the lateral distance between $D_H H$ and $D_M D$. The supply curve before a technological change (OS_0) shifts to (OS_1) after technological change. This results in a movement of the market equilibrium point from A to B. The quantity consumed increases from OQ_0 $(=q_0)$ to OQ_1 $(=q_1)$ consequent to a fall in price from OP_0 $(=p_0)$ to OP_1 $(=p_1)$. Consumers' surplus increases by the area $ACGB$. Producers' cash revenue changes from area $ACHQ_0$ to area $BGHQ_1$ with producers' home consumption remaining unchanged at OH. The cost of production changes from area AOQ_0 to BOQ_1.

Assuming that the real income value of home consumption of the product by producers is represented by the quantity consumed, income changes to producers are reflected in changes in their cash income. Whether producers' cash income (=revenue-cost) is increased by technical change depends on the demand and supply functions. In order to provide a formal mathematical treatment of the above relationships, assume a constant elasticity of demand function

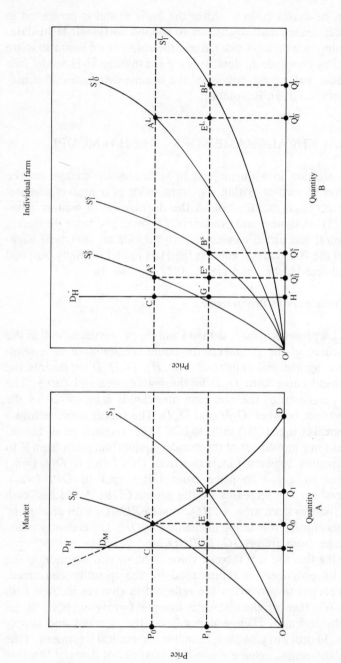

Figure 6.1 The distributional impact on market and individual farm of technological innovations in a semi-subsistence crop, after Hayami and Herdt (1977).

for the relevant range of the total demand function $D_H D_M D$

$$q = ap^{-\eta} \tag{6.1}$$

where p and q respectively represent price and quantity demanded of a subsistence crop, while income and other demand shifters are relegated to the constant term a and η is the price elasticity of demand.

Assume a constant elasticity of supply function

$$q = bp^{\beta} \tag{6.2}$$

p and q being price and quantity supplied; b includes supply shifters except technical change while β is the elasticity of supply. Assume that technological change leads to a k per cent shift in supply so that the supply function (OS_1) can be expressed as

$$q = b\,(1 + k)\,p^{\beta} \tag{6.3}$$

Employing Equations (6.1), (6.2) and (6.3) and using Taylor's expansion (see Thomas, 1968, pp.634–5), p_1 and q_1 can be approximated as follows:

$$p_1 \simeq p_0\,[1 - k/(\beta + \eta)] \tag{6.4}$$

and

$$q_1 \simeq q_0\,[1 - \eta k/(\beta + \eta)] \tag{6.5}$$

on the assumption that k is a relatively small percentage change.

Changes in consumers' surplus can be expressed as

$$\text{Area } ACGB = \text{Area } AP_0P_1B - \text{Area } CP_0P_1G$$

$$= \int_{p_1}^{p_0} ap^{-\eta}dp - q_0\,(1 - r)\,(p_0 - p_1) \simeq p_0q_0\,[kr/(\beta + \eta)] \tag{6.6}$$

r being the ratio of marketable surplus, i.e. HQ/OQ_0.

Change in producers' cash revenue is given by

$$\text{Area } BEQ_0Q_1 - \text{Area } ACGE$$

$$= p_1(q_1 - q_0) - q_0r(p_0 - p_1) \simeq p_0q_0k(\eta - r)/(\beta + \eta) \tag{6.7}$$

Equation (6.7) indicates that producers' cash revenue will increase only if $r < \eta$.

Cost of production will change by

$$\text{Area } BOQ_1 - \text{Area } AOQ_0$$

$$= [p_1q_1 - \int_0^{p_1} (1 + k) \, bp^\beta dp] - [p_0q_0 - \int_0^{p_0} bp^\beta dp]$$

$$\simeq p_0q_0k \, [\beta(\eta - 1)/(1 + \beta) \, (\beta + \eta)] \qquad (6.8)$$

Since $\eta < 1$, there will be a definite decline in cost. Consequently cash income of the producers will change by

change in cash revenue − change in cost

$$= p_0q_0k \, [(\eta - r)/(\beta + \eta)] - p_0q_0k \, [\beta(\eta - 1)/(1 + \beta) \, (\beta + \eta)]$$

$$\simeq p_0q_0k \, [(\eta - r) + \beta \, (1 - r)]/[(1 + \beta) \, (\beta + \eta)] \qquad (6.9)$$

6.2.2 Producers vs Producers

Figure 6.1B shows changes in equilibrium points of two types of individual producers corresponding to changes in market equilibrium in Figure 6.1A. The supply curves for small and large producers are represented by $O'S_0^s$ and $O'S_0^L$ respectively before the change in technology, and correspond to OS_0 in Figure 6.1A. The supply schedules of the small and large producers after the change in technology (OS_1) are represented by $O'S_1^s$ and $O'S_1^L$ respectively. Hayami and Herdt (1977, p.248) assume the quantity of home consumption for small and large producers to be identical.

As can be seen from Figure 6.1B, the equilibrium point of the small producer moves from A^s to B^s while that for the large producer moves from A^L to B^L. These movements lead to changes in cash revenues, cost of production and cash income of the producers for both groups of farmers. Following Hayami and Herdt, the same procedure as for changes in relative gains of consumers and producers (Equations (6.7) to (6.9)) is applied to derive the approximation formulae for analysing the impact of an aggregate supply shift by a factor of k on the ith producer. For the ith producer one has:

Change in cash revenue (ΔCR_i)
$$\simeq p_0 q_{oi} \left[k_i - k(\beta_i + r_i)/(\beta + \eta) \right] \qquad (6.10)$$

Change in cost (ΔC_i)
$$\simeq p_0 q_{oi} \left[(\beta_i)/(1 + b_i) \right] \left[k_i - k (1 + b_i)/(\beta + \eta) \right] \qquad (6.11)$$

Change in producers' income
$$\simeq p_0 q_{oi} \left[k - k_{ii}/(1 + \beta_i) - kr_i/(\beta + \eta) \right] \qquad (6.12)$$

q_{oi} and r_i respectively being the output and marketable surplus ratio of the ith producer prior to the introduction of the new technology. This gives $\beta = \Sigma w_i \beta_i$ and $k = \Sigma w_i k_i$, w_i being the share of the ith producer in the total output. One needs to mention that Equations (6.10) to (6.12) reduce to (6.7) to (6.9) if $k_i = k$ and $\beta_i = \beta$. Equation (6.12) indicates that the magnitude and direction of ΔI_i for the ith producer will be determined by two factors: (1) magnitudes of k_i and β_i relative to k and β; and (2) the magnitude of r_i. If r_i takes a smaller value ΔI_i takes a larger value.

6.2.3 Consumers vs Consumers

As presented in Equation (6.6), technical change leading to a fall in the price of a subsistence crop implies a clear gain in economic welfare (consumers' surplus) for non-producing households like urban consumers. Welfare gains, however, depend on the importance of foodgrains in their expenditure pattern. The percentage change in real income due to a fall in the foodgrain price can be approximated as

$$\Delta y/y = e \Delta p_f / p_f \qquad (6.13)$$

where $y = p_f q_f + p_{nf} q_{nf}$ and $e = p_f q_f / y$. Thus y is the total income expressed as the sum of expenses on food staples ($p_f q_f$) and other commodities ($p_{nf} q_{nf}$) and e is the proportion of income spent on foodgrains. The symbols p and q respectively represent price and quantity while subscripts f and nf symbolize food and non-food.

Since according to Equation (6.4), the percentage change in price due to a shift in supply function by k per cent shift is $k/(\beta + \eta)$, Equation (6.13) can be rewritten as:

$$\Delta y/y \simeq ek/(\beta + \eta) \tag{6.14}$$

As e is inversely related to per capita income, price decline flowing from technical change is likely to reduce gaps among urban consumers.

6.3   APPLICATION OF THE H-H MODEL TO BANGLADESHI DATA

It was noted in Chapter 2 that since the introduction of the new technology in the late 1960s, foodgrain yields per hectare of gross and net cropped land have increased considerably. The two measures of yield have increased by 22 and 28 per cent respectively between 1967–9 (average for the years 1967–8 to 1969–70) and 1982–4 (average for the years 1982–3 to 1984–5) (see Chapter 2, Table 2.2). The figures can be considered as surrogates for lower and upper bounds of foodgrain supply shifts (k). In this chapter, three alternative values, 0.15, 0.25 and 0.25 for k have been used.

Alamgir and Berlage (1973, p.396) estimated price elasticities of demand for foodgrains to be −0.172 and −0.177. R. Ahmed (1979, p.72) estimated a value of −0.19 for price elasticity of demand (η). The present study has used η values of −0.15, −0.20 and −0.25 with a view to evaluating the impacts of different price elasticities on the results. Based on a study by Cummings (1974), R. Ahmed (1978, p.128) estimated price elasticity of supply (β) to be 0.18. For the purpose of this study 0.15, 0.20, 0.25 and 0.30 as a set of possible values for β has been employed. R. Ahmed (1978, p.126) used a figure of 0.29 as ratio of marketed surplus (r). R. Ahmed (1981, pp.39–41) reports a marketable surplus ratio (r) between 0.18 and 0.22. We have in this study used a range of values for r: 0.15, 0.20 and 0.30.

The results of an exercise based on Equations (6.6) to (6.9) and the values of the specified parameters are set out in Table 6.1. It can be seen that consumers' surplus increases, owing to a decrease in price resulting from a supply shift. In all the cases considered, consumers stand to gain. A fall in price leads to a decline in producers' cash income in some cases but is outweighed by a decline in cost of production. However, consumers' gains seem to be higher in most cases and with a larger shift in supply and a greater degree of commercialization they tend to be even higher. Thus both producers and consumers are likely to gain from technological change in a

Table 6.1 Estimated percentage change in consumers' and producers' income from technological progress in foodgrain production: Bangladesh, 1967–9 to 1980–2, using the H-H method

Changes in	Percentage changes with specified parameters A: $k=0.15$, $r=0.15$					
	$\eta = 0.15$		$\eta = 0.18$		$\eta = 0.20$	
	$\beta = 0.15$	$\beta = 0.20$	$\beta = 0.15$	$\beta = 0.20$	$\beta = 0.15$	$\beta = 0.20$
Consumers' surplus	7.50	6.42	6.82	5.92	6.42	5.63
Producers' cash revenue	0.00	0.00	1.36	1.18	2.14	1.89
Production cost	−5.54	−6.07	−4.86	−5.39	−4.47	−5.00
Producers' cash income	5.54	6.07	6.22	6.57	6.59	6.89

	Percentage changes with specified parameters B: $k=0.20$, $r=0.25$				C: $k=0.25$, $r=0.30$
	$\eta = 0.20$		$\eta = 0.25$		$\eta = 0.25$
	$\beta = 0.20$	$\beta = 0.25$	$\beta = 0.20$	$\beta = 0.25$	$\beta = 0.30$
Consumers' surplus	12.50	11.11	11.11	10.00	13.64
Producers' cash revenue	−2.50	−2.22	0.00	0.00	−2.27
Production cost	−6.67	7.11	−5.56	−5.00	−7.87
Producers' cash income	4.17	4.89	5.56	5.00	5.60

subsistence agriculture like that of Bangladesh.

In order to analyse the impact of the new technology on the distribution of income among producers, one needs to compare changes in: (1) cash revenue, (2) cost of production and income of large and small farmers resulting from a change in price following the supply shift. Such comparisons are to be based on Equations (6.10) to (6.12).

For deriving results from Equations (6.10) to (6.12), one needs to specify plausible values of the parameters k, β and r. Some empirical evidence (e.g. Jones, 1984) suggests that there is little difference in the adoption of HYV technology among different classes of farmers. On that basis, it may seem reasonable to assume k_i to be the same for both groups of farmers. But the findings of the present study as reported in Chapter 6 and those of others (e.g. I. Ahmed, 1981; Asaduzzaman, 1979) indicate, differential adoption rates, both cross-sectionally and inter-temporally, cannot be ruled out. It was, therefore, thought essential to test the effect of differential rates of supply shifts for two categories of farmers. There does not seem to be any previous estimate of price elasticity of supply for different classes of producers. However, in the short run it is unlikely that β would be different for different classes of producers. The long-run situation is likely to be different. With greater command over resources, e.g. capital and credit, larger farmers are likely to be more responsive to price changes. It is, therefore, reasonable to assume a higher value of β for large farmers. The main difference between larger and small farmers lies primarily in respect of the proportion of marketable surplus (r). R. Ahmed (1981, p.41) provides the results of a survey which indicates that large farmers sell around three-quarters of their produce in the market while smaller farmers (including medium farmers) have a marketable surplus of around a quarter of their produce.

In the light of the above arguments, estimates of changes in cash revenue, cost of production and income of the two classes of producers are now presented in Table 6.2. Assume aggregate supply shift (k) to be 0.20 and the price elasticities of demand (η) and supply (β) to be 0.25 and 0.30 respectively. Also assume the following sets of alternative values for k_i, β_1, η_i and r_1 in order to derive the corresponding estimates:

Case 1: $k_1 = k_s = 0.20$ $\beta_1 = \beta_s = \beta = 0.30$
           $r_1 = 0.80, r_s = 0.20$

Case 2: $k_1 = 0.25, k_s = 0.15$ $\beta_1 = 0.40, \beta_s = 0.20$
           $r_1 = 0.85, r_s = 0.15$

Case 3: $k_1 = 0.30, k_s = 0.10$ $\beta_1 = 0.45, \beta_s = 0.15$
           $r_1 = 0.90, r_s = 0.10$

Table 6.2 sets out the results of the exercise. The results clearly indicate a relative gain in favour of small farmers far in excess of larger farmers. Whereas large farmers' incomes show declines, the income position of the small farmers tends to improve even when the values of k_1 are much higher than those of k_s.

To appreciate the gains for non-producer consumers, it is necessary to derive estimates of Equation (6.14). The gains accrued to various urban consumer groups vary directly with the relative importance of foodgrains in their family budgets. The differential impacts of technical change in the real incomes of urban households have been

Table 6.2 Estimated differential impacts of technical progress in rice production on large and small producers: Bangladesh, 1967–9 to 1980–2, using the H-H method

| Farmer Category | Specified parameters $k = 0.20$ $\eta = 0.25$ $\beta = 0.30$ | | | Percentage changes in | | |
	r_i	β_i	k_i	Cash Revenue	Production Cost	Cash Income
Case 1						
Small farmers	0.20	0.30	0.20	1.82	−6.29	8.11
Large farmers	0.80	0.30	0.20	−20.00	−6.29	−13.71
Case 2						
Small farmers	0.15	0.20	0.15	2.27	−4.77	7.04
Large farmers	0.85	0.40	0.25	−20.45	−8.83	−11.62

calculated using data from BBS (1980b, p. 20; 1984b, pp.709, 717) to Equation (6.14) with $\beta = 0.30$ and $\eta = 0.25$. A 20 per cent shift in the aggregate supply function is likely to lead to an 11 per cent increase in the real income of the consumer group with a monthly income of 300 taka. On the other hand, the same aggregate supply shift increases the real income of those in the monthly income bracket of over 2000 taka and over by 5 per cent. This indicates that the increase is relatively larger for those in the lower income bracket. These benefits are likely to accrue to urban consumers and landless labourers for whom foodgrains occupy the lion's share of the household expenditure.

6.4 LIMITATIONS OF THE H-H MODEL

It should be pointed out that the H-H model presented above is subject to a number of limitations. While the model has some value, there is a risk of drawing unwarranted assessments and policy conclusions from it.

First of all, technological change can cause a supply curve to shift in diverse ways, and the supply curve may not be of the mathematical form assumed in the H-H model. Hayami and Herdt (1977) consider only one type of shift. The nature of the supply shift can significantly influence the distribution between consumers and producers. This has been widely discussed in the recent literature. Duncan and Tisdell (1971) demonstrated that the nature of the supply shift is a critical determinant of benefits between producers and consumers, and more recent studies by Lindner and Jarrett (1978), Lund *et al.* (1980), Taylor (1980) and Wise (1978; 1981) have underlined this.

Secondly, the analysis is excessively partial. In consequence, for example, a loss of producers' surplus may appear to be the case from the single product model, but in reality no major loss of producers may occur. For instance, a movement from OS_0 to OS_1 reduces producers' surplus from the crop under consideration. However, yield-increasing technology may mean that *less land* has to be used for the production of same quantity of the crop. This will enable the the grower to use more land for the output of another crop or crops. Thus the producers' surplus overall (given that production is mixed) may not fall to the extent indicated in Figure 6.1. One can conceive of cases where producers' surplus goes up, since resources are released that are used to increase supplies of other crops. The outcome depends on the substitutability of one crop for another in the cultivation process.

Thirdly, the system is not closed. For example, if the income per head rises as a result of technological change, its impact on population growth is not predicted. Population is an exogenous variable in their model, unlike the Malthusian or Richardian models. It could well be endogenous. The possibility that technological change could, in the case of an *important* subsistence crop, increase income and population and *shift* the demand curve is not considered. Take the equation $D = n + \varepsilon g$ (see Johnston and Mellor, 1961, p.562), where D, n and g are respectively annual rates of growth in demand for food, of population and in per capita income while σ is the income elasticity of demand for food. The second term on the right-hand side is likely to be technology-induced, whereas the growth of population could contain elements of exogenous as well as endogenous elements.

Fourthly, it is unlikely that in reality the supply curve would pass through the origin. The relationship implies some supply at near zero prices. Some positive price is likely to be required to ensure supply to the market. This becomes important when areas above the supply curve are used for estimating variations in producers' surplus. Also the supply curves of cereals by larger producers are unlikely to be related to those of the smaller producers in the theoretical simple way assumed by Hayami and Herdt (1977). Their assumption has no empirical support.

Fifthly, there seems to be some implication in the writings of Hayami and Herdt that the more important the market element is,

the greater are the gains from technological change. But this over-
looks the possibility that production may become more specialized
and market oriented as the Green Revolution technology becomes
established. Farmers may become more dependent on purchased
inputs and require a constant stream of cash to purchase these. The
risks associated with these become a major influence on producers.
The technology may 'lock' them into a market system and their
subsistence demand may shift leftward.

Sixthly, the H-H model does not consider the question of variability of production which can also have influence on welfare. If
technological changes lead to greater variability in production and
hence supply of foodgrains, their prices may become more unstable.
In the LDCs this can have important welfare consequences for
low-income earners, and increase fluctuations in income received by
grain producers (Mellor, 1978). Furthermore, the question of sustainability is an important aspect of technological change which has
implications for income distribution (Douglass, 1984). There is concern that the Green Revolution may have resulted in a reduction in
genetic diversity (see, for example, Biggs and Clay, 1981; see also
Chapters 12–13). These global questions can be all too easily ignored
when one focuses on the analysis used by Hayami and Herdt.

There are, however, further important limitations of the H-H
model as far as income distribution is concerned in Bangladesh.
These are taken up in the next section.

6.5 FURTHER DISTRIBUTIONAL CONSEQUENCES OF NEW AGRICULTURAL TECHNOLOGY IN BANGLADESH

Hayami and Herdt (p.256) emphasize that their model

> abstracts from possible changes in the factor shares and factor
> ownership that might occur either as a result of technological
> change or at the same time for independent reasons. The final
> impact of such changes on the distribution would reflect the net
> effect of the new factor shares and factor ownership distribution as
> well as the real income effects discussed above.

The objective of this section is to consider some of these aspects in
the Bangladeshi context.

Table 6.3 sets out the relative shares of various factors of production in the output of traditional and modern varieties of rice in

Table 6.3   Relative share of labour and other factor inputs in total output per hectare of traditional and modern varieties of *aman* and *boro* rice: Bangladesh, 1980–1

| | Rice crop and variety | | | | | | | |
| | *Boro* (Bangladesh)[1] | | *Aman* (Thakurgaon) | | *Aman* (Rajshahi) | | *Aman* (Joydevpur)[2] | |
Input/Category	TRAD	MOD	TRAD	MOD	TRAD	MOD	TRAD	MOD
Material input from agriculture[a]	0.226	0.106	0.141	0.086	0.168	0.135	0.269	0.150
Material input from non-agriculture[b]	0.250	0.174	0.023	0.075	0.025	0.053	0.137	0.099
Return to human labour								
Total	0.476	0.288	0.218	0.196	0.280	0.322	0.254	0.174
Family	0.283	0.159	0.167	0.145	0.139	0.192	0.151	0.084
Hired	0.194	0.130	0.051	0.051	0.141	0.130	0.103	0.090
Return to capital[c]	0.048	0.431	0.618	0.643	0.527	0.490	0.340	0.516
Gross output	1.000	1.000	1.000	1.000	1.000	1.000	1.000	1.000
	5760	11460	4917	7890	6220	6666	7122	12646

Notes: TRAD and MOD respectively refer to traditional and modern varieties. Figures in the last row are values of gross returns per hectare in takas.

[1] Based on a survey from 12 districts in Bangladesh.
[2] For 1981–2.
[a] Including expenditure on seed and animal labour.
[b] Including expenditure on chemical fertilizer, pesticides and irrigation water.
[c] Including rent of land, interest on capital and net profit.

Source: Based on data from BMAF (1981b, pp.31, 36); M.M. Hossain and M.E. Harun (1983, pp.12–4); M.M. Hossain *et al.* (1981, pp.9, 14) and M.M. Hossain *et al.* (1982, pp.13–4).

Bangladesh. *Aman* rice has been taken as a proxy for rainfed *kharif* (wet) season rice crop, and *boro* rice for the irrigated foodgrains during the *rabi* (dry) season. Assuming prices to be the same for different groups of farmers, the differences in relative factor shares can be attributed to technological change. A few pertinent points emerge from the information presented in Table 6.3.

First, compared to modern varieties, the traditional varieties use a higher percentage of material inputs from within the agricultural complex. Secondly, the relative share of labour in the total output per hectare is much higher for traditional compared to modern rice varieties in all areas except Rajshahi. The difference is much more striking in case of irrigated rice. Similar differences can be noticed in respect of returns to capital. Thirdly, traditional varieties have much higher returns to capital in some places than in others. The difference in returns to capital between Joydevpur and Rajshahi or Thakurgaon is due to: (1) higher yields and (2) higher rice prices in 1981–2 compared to 1980–1. It may be that Joydevpur, being the centre for the Bangladesh Rice Research Institute (BRRI), has benefited more from rice research than either of the other two places which are far away from the BRRI headquarters.

Despite a decline in the relative shares of labour in modern rice compared to traditional varieties, the absolute income of labour has improved significantly during the *rabi* season. A recent study by Alauddin and Tisdell (1989c) indicates that employment during the *rabi* season has increased quite considerably. Without the introduction of the new technology, unemployment would most likely have increased. On the other hand, even though the relative share of labour for rainfed HYVs is not significantly lower than for their traditional counterparts, one needs to be reminded that the cultivation of the HYVs of rice during the *kharif* season does have very little impact on the demand for labour. Rainfed HYVs of rice have virtually the same labour requirement per hectare as that of the traditional varieties. Therefore the replacement of the traditional varieties by rainfed HYVs adds little to the overall demand for labour during the *kharif* season. Furthermore, the shift in hectarage from a more labour-intensive crop, jute, to different varieties of rice in recent years has had a depressing effect on the demand for labour (Alamgir, 1980, p.346; Alauddin and Mujeri, 1985, p.63; Alauddin and Tisdell, 1989c).

From Table 6.3 it can be seen that the relative share of family labour in total output is higher than that of hired labourers who come from the landless, near-landless or the dispossessed classes. This seems to support the hypothesis that employment has increased more in terms of the demand for family labour and less so in terms of hired labour (I. Ahmed, 1981). This may have led to the reduction in underemployment rather than unemployment *per se*. This has implications for income disparities between landowners with family labour and the landless and near-landless who usually work as wage labourers (BBS, 1981; 1986b; Cain, 1983; A.R. Khan, 1984).

Significant variations occur in the returns to family labour depending on the mode of operation (e.g. owner-operator or sharecropper)

Table 6.4   Returns to family labour for owner-cultivator and sharecropper for traditional and modern varieties of *boro* and *aman* rice: Bangladesh, 1980–1 and 1981–2

| | Boro paddy: Bangladesh | | | |
| | (a) Owner cultivator | | (b) Sharecropper | |
Item	Traditional	Modern	Traditional	Modern
Net income (taka/ha, cash cost basis)	2464.00	7329.00	–420.00	1599.00
Family labour applied (man-days/ha)	71.58	94.42	71.58	94.42
Returns to family labour (taka/day)	34.35	76.76	–5.87	16.93
Wage rate (taka/day)	15.25	15.73	15.25	15.73
	Aman paddy: Thakurgaon			
Net income (taka/ha, cash cost basis)	4376.00	6633.00	1917.00	2718.00
Family labour applied (man-days/ha)	83.80	102.60	83.80	102.60
Returns to family labour (taka/day)	53.20	64.65	22.28	26.49
Wage rate (taka/day)	10.59	10.59	10.59	10.59
	Aman paddy: Rajshahi			
Net income (taka/ha, cash cost basis)	4877.00	4703.00	1767.00	1370.00
Family labour applied (man-days/ha)	64.60	63.90	64.60	63.90
Returns to family labour (taka/day)	75.50	73.60	27.35	21.44
Wage rate (taka/day)	13.57	13.57	13.57	13.57
	Aman paddy: Joydevpur			
Net income (taka/ha, cash cost basis)	4045.00	8542.00	484.00	2219.00
Family labour applied (man-days/ha)	53.94	53.32	53.94	53.32
Returns to family labour (taka/day)	74.99	160.20	8.97	41.62
Wage rate (taka/day)	19.98	19.98	19.98	19.98

Note: The figures for Bangladesh, Thakurgaon and Rajshahi relate to 1980–1 and those for Joydevpur relate to 1981–2. The Bangladesh figures are based on a survey from 12 districts.

Source: Based on information contained in BMAF (1981b, p.38); M.M. Hossain and M.E. Harun (1983, p.12); M.M. Hossain *et al.* (1981, p.9); M.M. Hossain *et al.* (1982, p.13).

and technology (e.g. traditional or modern) as can be seen from Table 6.4. First of all, returns to family labour for the owner-operator are far in excess of those for the sharecropper. Secondly, for irrigated varieties, returns are much higher for either of the two groups of farmers. Thirdly, while for Joydevpur the returns to family labour (either owner or sharecropper) for the modern *aman* variety are much higher than those for the traditional *aman* variety, there is little difference between those for the two varieties in the Thakurgaon and Rajshahi areas. Fourthly, returns to family labour for sharecroppers cultivating the traditional varieties are in some cases higher than the market wage rate and lower in some others. One also observes significantly higher returns per hectare on a cash cost basis) for the owners compared to the sharecroppers. Thus the gap between the owner-cultivator and the sharecroppers is likely to widen further with more widespread diffusion of the new technology.

Irrigation is a critical input in the modern technological package. The ownership and control of this component of the new technology has had a significant impact on income distribution. First, smaller farmers have very little control over the ownership of this vital input. Secondly, through the patron–client relationship, the larger farmers who own irrigation equipment gain substantial revenues as rents for irrigation water sold to smaller farmers. Even when the ownership of irrigation equipment is cooperative, the smallholder is usually a very unequal partner. The siting of tubewells normally takes place on the large farmer's plot, giving him control and readier access to irragation water (Alam, 1977). This seems to be supported by farm-level evidence from Ekdala, one of the survey villages reported in Chapter 7.

Consequent upon inequality in the distribution of land (BBS, 1981; 1986b), farmers have unequal access to social, political and economic power critical to the decision-making process underlying the allocation of resources to promote agricultural development. In Chapter 6 we provided farm-level evidence that larger farm households were innovators in adopting the new technology, even though smaller farms eventually caught up with them. Hayami and Ruttan (1984, pp. 53–4) point out that early adopters may capture large excess profits (Schumpeterian entrepreneurial profits) from the use of innovations 'without forcing down the product prices or bidding up factor prices appreciably'. But with wide diffusion of the new technology, 'innovators' excess profit will be lost as product and factor prices move towards a new equilibrium. In the long run, the relative share of labour will return to the same level as before the introduction of MV if MV represents a neutral technological change'.

However, the institutions that affect the adoption and diffusion of technological innovations are biased towards those farmers who are better endowed with land resources (see for example, A.R. Khan, 1979). Thus the apparent 'scale-neutrality' of the 'Green Revolution' technologies may not be meaningfully manifested in the actual process of agricultural development. Based on recent Indian experience, Diwan and Kallianpur (1986) raise doubts about the scale-neutrality of biological technology inputs. They claim that the 'scale-neutrality hypothesis follows from the agronomy studies conducted on experimental farms. By abstracting from socio-economic relations these studies may not be able to capture some of the implied biases of the biological technology inputs' (p.127; see also Feder *et al.*, 1985, p.288).

6.6 CONCLUDING OBSERVATIONS

The application of the H-H model to determine gains from modern varieties of rice in Bangladesh and the distribution of these gains between consumers and producers indicates that consumers' surplus is much greater than it would have been if the HYVs had not been introduced. These findings are consistent with those of Hayami and Herdt (1977, pp.250–1) for the Philippines rice economy. The modern varieties, by keeping the real price lower than otherwise would have been the case, have tended to be income-equalizing for urban consumers. Nevertheless it masks the more general result which we have outlined in Chapter 5 of a failure of real incomes in Bangladesh to rise with the 'Green Revolution' even though they clearly would have risen in the absence of population growth.

The impact of modern varieties of rice on incomes of Bangladeshi farmers and the distribution of income between those involved in production is more complex. The H-H partial model suggests that, given the relatively inelastic demand for rice in Bangladesh, the real cash income of producers has risen slightly as a result of the new technologies. But it is suggested that if a less partial view is taken and if account is taken of the lower cost of obtaining home-consumed produce, the increase in income may be greater. In any event, there are dangers in using such a partial model to predict the developmental consequences of technological changes affecting a staple crop. Attention also needs to be given to the possibility that the supply curve may not have the simple form and pivot in the way supposed by Hayami and Herdt.

While the H-H model is simple to apply, it is best used as first approximation or starting-point rather than a final solution. Furthermore, the H-H model does not deal specifically with changes in factor shares in farm production. It is pointed out here that the adoption of HYVs has led to important variations in factor shares in Bangladeshi rice production. For instance, the relative share of labour has fallen. However, the absolute share has increased and it seems that rural employment has risen as a result of the new technologies.

A fuller appreciation of distributional consequences requires account to be taken of the distribution of land and other resources that affect rural income. With greater population pressure and penetration of technological and market forces, the distribution of these resources changes over time and in turn affects the welfare of the rural population. These and other relevant issues are taken up in Chapter 8. But they are affected by the pattern of adoption and diffusion of HYVs of crops and other associated technology at the farm-level and this is discussed in Chapter 7.

7 Patterns and Determinants of Farm-Level Adoption and Diffusion of High-Yielding Varieties (HYVs) and their Technology

7.1 INTRODUCTION

The question of adoption and diffusion of innovations in general, and agricultural innovations in particular, has attracted considerable attention in recent years. The volume of published research on patterns and determinants of adoption behaviour both at the theoretical and empirical levels is vast and growing rapidly.[1] This emphasis appears justified because of the large potential economic welfare implications of technological change (Mellor, 1985). Much of the argument about the existence of differential gains seems to have resulted from the evidence of differential rates of adoption and diffusion of the new agricultural technology.

This chapter examines the process of adoption of HYV technology in Bangladesh, employing farm-level data from two Bangladeshi villages. It is the first time that a study of this kind has been undertaken for Bangladesh. Particular attention is paid to the behaviour of small farms compared with medium and large farms. The existing literature on adoption and diffusion has concentrated primarily on identifying classes of innovators and imitators and the factors that distinguish them, and insufficient attention has been given to welfare significance and the distributional consequences the dynamics of adoption is likely to entail (Feder and O'Mara, 1981, excepted).

7.2 ADOPTION AND DIFFUSION OF AGRICULTURAL INNOVATIONS: A BRIEF REVIEW OF THE LITERATURE

A number of models have been developed to explain adoption of innovations both by individual farmers and at the aggregate level. Aggregate adoption models (Bera and Kelley, 1990) investigate the dynamic behaviour of the diffusion process over time and provide empirical parameter estimates for the S-shaped diffusion path (Griliches, 1957; Mansfield, 1961; 1968). Some recent studies (e.g. Feder and O'Mara, 1981; 1982; Feder and Slade, 1984) investigate the dynamics of the adoption process, in particular, the pre-adoption phase. A feature of these studies is the importance of acquisition of information (Shetty, 1968; Rogers and Shoemaker, 1971; Lindner *et al.*, 1979; Feder and Slade, 1984). Feder and Slade (p.320) argue that before adoption of an innovation takes place a 'certain critical level of cumulative information must be attained'. They suggest that operators of larger farms with better access to information, or farms with more human capital, are likely to adopt earlier than others. Furthermore, during the initial stages of the diffusion process, operators of larger farms are inclined to allocate more resources to the acquisition of information.[2]

A number of empirical studies employing farm-level data have considered the issues of adoption and diffusion of agricultural innovations in various parts of the world. Hayami and Ruttan (1984, pp. 48–9), after reviewing evidence for several Asian and African countries, concluded that 'the available evidence indicates that neither farm size nor tenure has been a serious constraint on the MV [modern variety] adoption. . . .On the average, small farmers adopted the MV technology even more rapidly than large farmers.' They claim that where evidence to the contrary has been found, it seems to be 'an exception rather than a norm'. However, the Hayami and Ruttan contention seems inconsistent with that of Ruttan (1977, p.22) who presents a 'stylized model of HYV diffusion process'. Ruttan argues that while the smaller farmers initially lag behind larger ones in adopting HYV, they eventually catch up.

B. Sen (1974, p.104), summing up the Indian experience, claims that 'a broad spectrum of farms in which the small and the medium sized farms predominate, are using the new varieties. . .[due] partly to the inherited pattern of distribution of irrigated land'. Furthermore, even though in some areas large farms may have led the adoption process, smaller farms have quickly followed, whereas in some other areas the small and medium farms have taken the lead. Palmer (1976) presents evidence from the Far East that long-run adoption rates of large farms tended to be higher than those of the smaller farms, even though this was not initially so.

Muthia (1971) provides testimony from Indian experience that smaller farms initially lag behind large farms in adoption but rapidly catch up. Shetty (1968, p.1273), using data from two South Indian villages, concludes than 'it is the larger, wealthier more literate and younger ones who lead the innovation'. In Shetty's view the process of adoption is a more 'gradual sequence through innovators-imitators-laggards'. Shetty (p.1281) further claims that 'variation in time of adoption is mainly due to differences among farmers in relation to access to information and supply'.

Barker and Herdt (1978, pp.90–1) show that the smaller farms in their sample adopted HYVs before the medium and large farms. However, their study revealed no significant difference in the time-pattern of adoption when the villages were grouped on the basis of the degree of inequality of land distribution (measured by Gini coefficient) even though the larger farms appeared to be early adopters (p.94). But in one village which had the highest degree of inequality in the distribution of land (Gini coefficient = 0.56), the difference between small and large farms was striking (see also Parthasarathy and Prasad, 1978). On the basis of this and other evidence, Barker and Herdt (p.98) argue that the 'notion that "the rich get richer" and "the poor get poorer" is questionable.' They recognize, however, that 'the benefits from the new technology were clearly associated with farm-size in some villages. . .institutional differences in rural and farm organisation influence the relationship between the introduction of new technology and farm-size'. However, Barker and Herdt do not provide any explanation of the observed time-pattern of adoption. A more recent study by Herdt (1987, p.337), based on surveys in the Philippines, supports 'the hypothesis that small farmers adopted the technology somewhat more slowly in the central Luzon sample, but they caught up to the larger farmers after some time.'

Thus most of the evidence reviewed above indicates a positive association between farm size and HYV adoption. As Feder *et al.* (1985, p.272) point out, 'even seemingly neutral technologies such as the HYVs may entail significant setup costs in terms of learning, locating and developing markets, and training hired labor. When these factors are considered as fixed expenses, the theoretical models imply that they tend to discourage adoption by small farmers'.

A number of recent studies have addressed the question of adoption and diffusion of new agricultural technology in Bangladeshi agriculture. Asaduzzaman (1979), I. Ahmed (1981), Atiqur Rhaman (1981), R.I. Rahman (1983) and Jones (1984) have used farm-level data from different areas of Bangladesh to study the adoption of the HYV (high-yielding variety) technology and the factors underlying any emerging pattern. Atiqur Rahman (1981), using data from Mymensingh and Comilla, examined adoption of HYVs in general, in that he did not distinguish between seasons. Atiqur Rahman argued that resource-constraints on various classes of farmers affected that adoption decisions. Asaduzzaman (1979) collected data in Rangpur and Noakhali and examined adoption of HYVs in the *aman* season. Asaduzzaman's study (p.33) found that (on operational basis) while a higher percentage of larger farmers adopted HYVs, the smaller farmers among the adopters allocated a higher percentage of farm area to HYV cultivation (p.40). I. Ahmed (1981, p.23–6) distinguished between two seasons, *aman* and *boro*, and using data from Sylhet, Noakhali and Bogra found (on an ownership basis) that a higher percentage of larger farmers adopted HYVs while the per cent area allocated to HYVs among the adopters was negatively associated with farm size. On both counts owner-farmers had higher adoption rates compared to tenant farmers. One of the important factors that determined adoption rate as reported by I. Ahmed (1981) was the extent of irrigation. The village with a higher percentage of area under irrigation was found to be adopting HYVs at a higher rate. R.I. Rahman (1983) used data from Dhaka district and emphasized the role of supply-side factors in determining the adoption of HYVs. These included, among other things, the supply of irrigation water and agricultural credit. Furthermore, R.I. Rahman reported (p.67) an adoption pattern by farm size similar to the one by I. Ahmed (1981).

Jones (1984), on the other hand, using village level data from the Dhaka district of Bangladesh, found some evidence to the contrary. Disaggregating by ownership pattern and season, Jones found a U-shaped adoption pattern. In Jones's study the smallest farmers had the largest proportion of (owner-cultivated) land under HYVs, followed by the larger farmers, while the medium farmers had the smallest proportion of such land devoted to the new technology (p.202). To quote Jones (p.203), 'smaller farmers then are not the lower adopters of HYVs than larger farmers. Rather it is the smallest farmers,. . .who are the highest and fastest adopters of the new technology.' However, Jones notes that the U-shaped relationship is a dynamic one and showed some changes between 1978 and 1980 in that large farmers appeared to be as high adopters as the small farmers if not higher (Jones, 1984, Table 10.4).

These studies of Bangladesh agriculture suffer from methodological limitations. While Atiqur Rahman (1981) and R.I. Rahman (1983) provide adequate circumstantial evidence of the factors underlying differential adoption, they do not carry out further statistical testing to examine the statistical significance of the strength of the causality relationships. Atiqur Rahman (1981) suffers from further limitation in that HYV adoption is not disaggregated by season. Among other things, the risk factor seems to differ between rainfed and irrigated crops (I. Ahmed 1981, p.23). Jones (1984), while disaggregating HYV adoption by seasons, does not analyse the factors underlying the observed adoption pattern. Moreover, no further statistical tests are undertaken to provide any adequate explanation of the process of adoption of the HYV technology. Both Asaduzzaman (1979) and I. Ahmed (1981) subject their data to further statistical analysis to provide a more in-depth analysis of the adoption process. However, Asaduzzaman is concerned only with the rainfed crops, and therefore leaves no scope for comparison between seasons. The extent to which irrigation, perhaps the most critical factor in the expansion of HYV area, helps explain differential adoption rates cannot be ascertained. (I. Ahmed, 1981) is methodologically superior to Asaduzzaman (1979) in this respect. However, I. Ahmed suffers from the limitation that the statistical analysis is carried out in terms of pooled data, even though village dummies are employed to account for regional differences. In our view, the process of adoption would have been better highlighted if data were analysed separately for each village. This would have provided a better analytical and comparative basis of within and between village adoption process.

Apart from the methodological issues discussed above, some of the studies (e.g. I. Ahmed, 1981; Asaduzzaman, 1979; Atiqur Rahman, 1981) employ information which dates back to the early or mid-1970s, while others use data relating to 1978 and 1980 (e.g. Jones, 1984; R.I. Rahman, 1983). Some changes have taken place in Bangladesh agriculture since these studies have been completed. For instance, one of the key elements of growth in Bangladeshi food production in the post-Green Revolution period is the increased intensity of cropping. In the last few years this seems to have stabilized just over the 150 per cent mark for Bangladesh (Alauddin and Tisdell, 1987). In view of this and other changes, new studies employing more recent data are warranted.

7.3 ADOPTION OF HYV TECHNOLOGY: ISSUES AND HYPOTHESES

Various factors may affect farmers' decision to adopt an innovation. These include, among others, farm size, tenurial status, membership of farmers' organizations, level of education, access to critical inputs like irrigation and credit, and subsistence pressure. Other factors, like objective and subjective riskiness of the innovation and farmers' perceptions of the profitability and expected increase in income, also affect adoption decision. This section has two objectives. First, it provides various indicators of adoption. Secondly, it presents a brief description of the theoretical and conceptual framework and sets forth the hypotheses that are to be empirically investigated later in the chapter.

7.3.1 Indicators of HYV Adoption

Following I. Ahmed (1981) and Lipton (1978), we consider four indicators of adoption as follows:

(a) Crude adoption rate: Defined as the ratio of the number of farmers cultivating HYVs to the total number of farmers.
(b) Intensity of adoption: Defined as the percentage of farm area under HYV.
(c) Index of participation: Defined as the product of the crude adoption rate and intensity of adoption.
(d) Propensity to adopt: Defined as the likelihood of a farmer adopting the HYV innovation.

7.3.2 Adoption and Other Variables: *A Priori* Considerations

Agricultural production in Bangladesh is organized around small family farms with fragmented plots. Socio-economic factors apart, an average Bangladeshi peasant confronts extreme natural constraints imposed by topographic and climatic conditions. Cultivation practices are still basically traditional, even though the introduction of new agricultural technology has made steady progress in the last two decades. Average family size is well above five, indicating a highly unfavourable land–man ratio, given that the average size of holding is small. Under such a scenario, survival and food consumption seem to be the only major concern of an average Bangladeshi peasant household.

The analysis of peasant behaviour towards adoption of innovation can be facilitated by referring to the Chayanovian (see Thorner *et al.*, 1966) and safety-first models.[3] In the former model, requirement for absolute subsistence (total consumption need), which increases with the growth in family size, is the critical determinant of a peasant family's economic activity. A peasant household in such a model is assumed to respond to growing absolute subsistence by, among other things, a greater acquisition of the means of production, primarily land, either by its purchase or by extension of margin. In the safety-first models, a farm household is assumed to ensure survival for itself and, therefore, it wants to avoid the risk of its income or return falling below certain minimum (subsistence) level (Roy, 1952; Shahabuddin *et al.*, 1986; Tisdell, 1962). How the absolute and relative subsistence requirements and other variables are likely to influence the attitude of an average Bangladeshi peasant toward adoption of HYV innovation, has been discussed in detail by Alauddin and Tisdell (1988e, pp. 186–90) so we do not wish to repeat this discussion here. Instead Table 7.1 summarizes the hypotheses that are being tested and sets out the definitions of the relevant variables.

Table 7.1   Hypotheses and relevant variables: Farm-level determinants of HYV technology adoption in Bangladesh

	Influence on adoption on a priori grounds			
Factors	*Crude rate of adoption*	*Itensity of adoption*	*Index of participation*	*Propensity to adopt*
Farm size	Positive	Negative	Positive	Positive
Tenancy	Positive	Positive	Positive	Positive
Absolute subsistence pressure	Positive	Negative	Unknown	Positive
Relative subsistence pressure	Negative	Positive	Unknown	Negative

Table 7.1 *continued*

| Factors | Influence on adoption on a priori grounds | | | |
	Crude rate of adoption	Itensity of adoption	Index of participation	Propensity to adopt
Agricultural worker	Positive	Positive	Positive	Positive
Labour scarcity/Land abundance	Positive	Negative	Unknown	Positive
Education	Positive	Positive	Positive	Positive
Irrigation	Positive	Positive	Positive	Positive

Definition of variables

Farm size:	Amount of owned or operated land (OWNAREA or OPERA).
Subsistence pressure:	Absolute subsistence pressure (ABSUB) is measured in terms of number of consuming units of male adult equivalents. Adults are defined as persons of 10 years and over. Female adults and children have been converted into male adult equivalents using conversion factors of 0.90 and 0.50 respectively (cf. Asaduzzaman, 1979, p.30). Relative subsistence pressure is defined as the ratio of absolute subsistence pressure to farm size (ABSUB/OWNAREA = SUBSIST or ABSUB/OPERA = SUBSIST1).
Agricultural worker:	Number of adult male family members available for agricultural work excluding full-time students (AGWORKER).
Labour scarcity:	Defined as the ratio of agricultural workers to size of owned land (LABSUP = AGWORKER/OWNAREA).
Education:	An educational score for each farm household has been defined on the basis of information on the level of education for each adult member of the household. For each level we have assigned an arbitrary score as follows: Above secondary = 1.00; above primary and up to secondary = 0.50; primary = 0.25. The aggregate of these scores is the educational score of the household (EDU). A zero score implies that all its adult members are illiterate.
Tenancy:	Operated land as a percentage of own land (PCOPERA).
Irrigated area:	Amount of irrigated land including rented-in land (IRRI). Percentage area irrigated implies irrigated land as a percentage of operated area (PCIRRI).

7.4   DESCRIPTION OF SURVEY METHOD AND SURVEY AREAS

The data used in this chapter are derived from sample surveys in two Bangladeshi villages. The collected data relate mainly to the crop year 1985–6. Employing a direct questionnaire method, we collected data at the farm level with the aid of research investigators. The fieldwork was conducted during the August–October period in 1986 (for further details see Alauddin, 1988). The survey villages of Ekdala in the north-western district of (greater) Rajshahi and South Rampur in the eastern district of Comilla were selected purposively. We chose them for three reasons: (*a*) their long tradition with HYV technology; (*b*) relatively easy access by road or train from the respective district headquarters and the capital city of the country; and (*c*) their geographic separation and location in different ecological zones.

Geophysically South Rampur belongs to the more frequently flooded and fertile areas of the eastern region of Bangladesh. The village experiences an average rainfall of well over 200 cm and is located in the high rainfall zone (BBS, 1985b, p.28). South Rampur is flooded more or less every year and is a flood-prone village in the Surma–Kusiyara flood plain. Ekdala, on the other hand, belongs to the low rainfall area and experiences an average annual rainfall of 120–150 cm (BBS, 1985b, p.28). The village is in the dry zone and can be considered drought prone, located in the lower Mohananda and higher part of the Ganges flood plains (BBS, 1985b, p.16). Apart from differences in geophysical characteristics, the two villages differ significantly from one another in terms of (*a*) pattern of land ownership and distribution; (*b*) intensity of irrigation; (*c*) cropping pattern and intensity of cropping and (*d*) incidence of landlessness (Alam, 1984; Saha, 1978).

The year 1985–6 was a fairly normal one for both the villages. It is also worth mentioning that geophysically both South Rampur and Ekdala may be considered to be somewhat typical of many villages in their respective ecological zones. Technologically, however, both the villages are fairly progressive compared to many villages in Bangladesh.

In all, 58 land-owning farm households were interviewed in each of the two villages. The samples constituted about 35 per cent and 43 per cent of the total land-owning households in Ekdala and South Rampur respectively. Following the latest agricultural census classification (BBS, 1986b, p.30; see also BBS, 1981) three farm categories

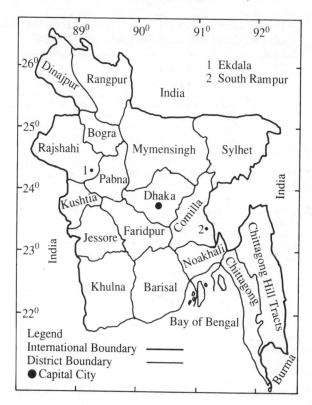

Figure 7.1   Map of Bangladesh showing districts and the two survey villages: (1) Ekdala (Rajshahi) and (2) South Rampur (Comilla)

for Ekdala were defined as: small farms (up to 1 hectare), medium farms (1–3 hectares) and large farms (3 hectares and above). The number of Ekdala farmers interviewed in each category were 40, 11 and 7 respectively, which corresponded to the proportion of each category in the total population of landowners in the village. In South Rampur a slightly different classification was employed, as there were rarely any large farmers according to the above classification (cf. Asaduzzaman, 1979, p.40). For South Rampur the three farm categories were defined as: (1) small (up to 1 hectare); (2) medium (1–2 hectares); and (3) large (2 hectares and above). The number of South Rampur farmers interviewed in small, medium and large farm categories were 35, 15 and 8 respectively.

Table 7.2   Broad pattern of HYV adoption in two Bangladesh villages: Ekdala (Rajshahi) and South Rampur (Comilla), 1985–6

(a) Land under HYV cultivation by season

	Ekdala			South Rampur		
	Own land	Sharecropped land	Total	Own land	Sharecropped land	Total
Rabi HYV (ha)	27.815	3.811	31.626	47.398	4.569	51.927
percentage of net cropped area	46.171	48.167	46.403	99.745	96.903	99.914
Kharif HYV (ha)	24.082	2.920	27.002	19.506	1.457	20.963
percentage of net cropped area	39.974	36.060	39.619	41.049	30.090	40.133
All HYVs (gross ha)	51.897	6.731	58.628	66.904	6.026	72.890
percentage of net cropped area	86.146	85.073	86.022	140.794	127.805	139.545

(b) Percentage of land under HYVs: Some further indicators

	Ekdala			South Rampur		
Indicator	Rabi HYV	Kharif HYV	All HYV	Rabi HYV	Kharif HYV	All HYV
HYV area as percentage of seasonal cereal area	100.000	52.048	66.623	100.000	40.573	70.334
Percentage of all HYV area	53.944	46.056	100.000	71.240	28.760	100.000

7.5   EMPIRICAL RESULTS: STATIC CROSS-SECTIONAL ANALYSIS

7.5.1   Broad Pattern

Table 7.2 provides a broad picture of the extent of HYV adoption in the two study villages. Significant differences can be noticed in regard to the adoption of *rabi* (dry) season cereals. Whereas all the *rabi* season rice crop is under HYV in South Rampur, only less than half of the net cropped area is allocated to *rabi* HYV cereals in Ekdala. If wheat is excluded, only 28 per cent of the net cropped area is under HYV *boro* rice. However, there is little or no difference in the intensity of adoption during the *kharif* (wet) season. In both the villages 40 per cent of the net cropped area is planted with *aman* HYV rice. When the gross area cropped with all HYVs is expressed as a percentage of net cropped area, the contrasting pattern comes into sharper focus. The percentage for South Rampur is more than 60 per cent higher than that of Ekdala. Also there is an inter-village difference in the relative share of *rabi* and *kharif* HYV areas in (gross) HYV area. For Ekdala, there is no significant difference between the relative shares of *rabi* and *kharif* HYV areas. However, for South Rampur, the relative share of *boro* HYV area is 2.5 times that of *aman* HYV area.

Information on crude adoption rate, intensity of adoption and index of participation in Ekdala and South Rampur is set out in Table 7.3. As rice is the dominant crop in Ekdala and it is the only crop in South Rampur, data on the adoption of HYV rice disaggregated by season and by farm size are presented. Several points emerge from a closer examination of the information contained in Table 7.3.

1. The crude adoption rate for HYV *boro* is lower among smaller farmers of Ekdala. It is the highest for the medium farmers, followed closely by the large farmers. For *aman* HYV it is systematically higher for larger farmers. In South Rampur crude adoption rate for *aman* HYV increases with farm size. All the non-adopters are from the small-farm category.
2. For Ekdala the intensity of adoption of *boro* HYV is lower for larger farmers. However, there does not seem to be any systematic relationship for *aman* HYVs in either village.
3. The index of participation follows a similar pattern to that of intensity for *boro* HYV in Ekdala. But in both areas for *aman* HYV it tends to rise with farm size, although not systematically.

7.5.2 Intensity of Adoption and Farm Size: A Simple Analysis

The objective of this section is to show how intensity of adoption of HYVs is related to the overall size of farm-holdings in our samples. We use farm-size as the dependent variable. I. Ahmed (1981) and Asaduzzaman (1979) contend that the intensity of adoption of HYVs tends to decline with farm size. However, Jones (1984) claims that intensity of adoption of HYVs tends to fall at first with increase in farm size and then rise, so that the relationship is U-shaped. Our results for Ekdala and South Rampur support the hypothesis of Ahmed and Asaduzzaman in cases where intensity of adoption varies with farm size. Our observations are, however, incompatible with Jones's hypothesis. In particular, there is no evidence whatsoever

Table 7.3 Various indicators of HYV adoption by farm size: Ekdala (Rajshahi) and South Rampur (Comilla), 1985–6

| | Indicators of adoption | | | | | | | |
| | Boro HYV rice | | | | Aman HYV rice | | | |
	Number of farmers	Crude rate of adoption	Intensity of adoption	Index of participation	Number of farmers	Crude rate of adoption	Intensity of adoption	Index of participation
(a) Ekdala								
Small	29	72.50	51.27	37.17	27	67.50	45.91	30.99
Medium	10	90.91	34.00	30.91	9	81.80	53.56	43.73
Large	6	85.71	22.10	18.94	7	100.00	40.69	40.69
All farms	45	77.59	34.24	25.10	43	74.14	45.12	33.43
(b) South Rampur								
Small	35	100.00	100.00	100.00	29	82.86	40.63	33.42
Medium	15	100.00	100.00	100.00	15	100.00	41.64	41.64
Large	8	100.00	100.00	100.00	8	100.00	37.93	37.93
All farms	58	100.00	100.00	100.00	58	89.67	40.13	35.98

that intensity of adoption of HYVs rises after a particular farm size is reached.

At this stage one might wonder if the differences between Jones (1984) and the present study in respect of the relationship between farm size and intensity of adoption may be attributed to differences in definition of 'small' and 'large' farms. In particular, questions might arise whether Jones considers those farms that are in a condition of 'immiserisation' while our observations have excluded them, so that our 'small' is Jones's 'medium'. A comparison of the classifications shows, despite some differences, that there is no fundamental differ- ence in the classifications. Consider Jones's (1984, p.203) classifi- cation of farm size (in hectares): 0–0.39, 0.40–0.79, 0.80–1.19, 1.20–1.59, and above 1.60. The first two groups constitute Jones's 'small' which, though not identical, is similar to ours (0–1.0 ha, see the preceding section). The third and fourth groups taken together make up Jones's 'medium' which contrasts with ours (1–2 ha for South Rampur and 1–3 ha for Ekdala). Despite some differences in classification, the groups in the two studies do overlap and the present study does not exclude the farms which are in condition of 'immiserisation'. For instance, our Ekdala sample of adopting farms includes five observations of below 0.25 h and five others between 0.25 and 0.33 h. This can also be seen from Figure 7.2. There are similar observations from the South Rampur sample.

Consider our results for intensity of adoption in South Rampur and Ekdala for the *rabi* season (INTNRHYV) and then for the *kharif* season (INTNKHYV). In South Rampur, every household adopts HYV on all its land in the *rabi* season. So

$$\text{INTNRHYV (South Rampur)} = 100 \qquad (7.1)$$

This case is not consistent with any of the hypotheses mentioned above. However, in Ekdala, intensity of *rabi* HYV adoption does vary according to farm size and broadly appears to decline with farm size as can be seen from Figure 7.2.

To estimate the relationship between intensity of adoption of HYV and farm size (OPERA) in Ekdala in the *rabi* season, we fitted a linear and a semi-log function by least squares to the scatter of observations. The results for the linear and semi-log functions are respectively:

$$\begin{aligned} \text{INTNRHYV} \qquad &\text{(Ekdala)} = 65.29 - 11.4480\text{PERA} \\ &(R^2 = 0.3390,\ t = 4.70) \qquad (7.2) \end{aligned}$$

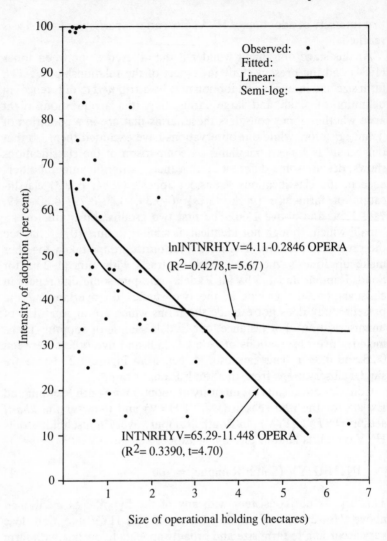

Figure 7.2 Relationship between intensity of adoption and (INTNRHYV) and the size of operational farm-holding (OPERA) for Ekdala sample in the *rabi* season

$$\text{lnINTNRHYV} \quad \text{(Ekdala)} = 45.64 - 0.1860\text{PERA}$$
$$(R^2 = 0.5011, t = 6.57) \tag{7.3}$$

While neither of these functions have strong explanatory power, the *t*-values for the coefficients are highly significant. The semi-log function gives a better fit than the linear one and indicates that intensity of adoption decreases at a decreasing rate with increase in the size of the operational holding.

For the *kharif* season a much weaker negative relationship seems to exist between the intensity of *kharif* HYV adoption (INTNKHYV) in both Ekdala and South Rampur. For the linear functions, the least squares fits were respectively:

$$\text{INTNKHYV} \quad \text{(Ekdala)} = 57.005 - 4.55260\text{PERA}$$
$$(R^2 = 0.0631, t = 1.64) \tag{7.4}$$

$$\text{INTNKHYV} \quad \text{(South Rampur)} = 43.904 - 2.11800\text{PERA}$$
$$(R^2 = 0.0146, t = 0.86) \tag{7.5}$$

In those cases where intensity of adoption of HYV varies with farm size, our empirical evidence tends to support the hypothesis of I. Ahmed (1981) and Asaduzzaman (1979) but not that of Jones (1984). However, it is also clear that farm size alone has low explanatory power. Clearly, additional factors to farm size need to be taken into account to model the situation accurately. The remainder of our analysis is designed to take these additional factors into consideration.

7.5.3 Results of Bivariate Analysis

In order to see the strength of the association between various measures of adoption on the one hand and the relevant determinants on the other we have applied the (Yates corrected) chi-square test. The results are set out in Table 7.4. The direction of association (positive or negative) is based on 2×2 contingency tables. For South Rampur, the following picture emerges. As expected, crude adoption rate is negatively associated with relative subsistence pressure. It seems to be positively associated with education, but not significant at the 5 per cent level. Adoption is negatively associated with labour supply which is contrary to expectations. This is possibly because of the smallness of the size of holdings: the farmers with abundant supply of labour may have to look for work outside the

134

Table 7.4 Bivariate analysis of association of various indicators of HYV adoption with different factors: Ekdala (Rajshahi) and South Rampur (Comilla), 1985–6

Factors	Crude adoption rate			Indicators of adoption Intensity of adoption			Index of participation		
	Chi-square	P-value	Relation	Chi-square	P-value	Relation	Chi-square	P-value	Relation
(a) Ekdala (boro HYV rice)									
Own area	1.392	0.2381	***	5.957	0.0147	Negative	6.817	0.0090	Positive
Operated area	2.489	0.1147	***	7.750	0.0054	Negative	3.125	0.0771	***
Irrigation	5.535	0.0186	Positive	24.851	0.0000	Positive ***	13.363	0.0003	Positive ***
Agricultural worker	0.586	0.4441	***	0.000	1.0000	***	0.384	0.5356	***
Education	0.316	0.5740	***	0.000	1.0000	***	1.548	0.2135	***
Labour scarcity	0.000	1.0000	***	3.695	0.0546	***	0.001	0.9757	***
Relative subsistence[a]	1.454	0.2279	***	6.297	0.0121	Positive	0.000	1.0000	***
Relative subsistence[b]	0.239	0.6249	***	6.591	0.0102	Positive	0.596	0.4401	***
Tenancy	0.000	1.0000	***	0.000	1.0000	***	1.949	0.1627	***
Absolute subsistence	0.000	0.9909	***	0.551	0.4578	***	2.423	0.1196	***
(b) Ekdala (aman HYV rice)									
Own area	2.379	0.1230	***	0.013	0.9104	***	15.430	0.0001	Positive
Operated area	3.886	0.0487	Positive	0.448	0.5033	***	6.369	0.0116	Positive
Irrigation	11.154	0.0008	Positive	18.258	0.0000	Positive	15.058	0.0001	Positive

Agricultural worker	0.000	1.0000	***	7.778	0.0053	Positive	1.891	0.1691	***
Education	2.716	0.0993	***	0.688	0.4069	***	6.461	0.0110	Positive
Labour scarcity	0.835	0.3608	***	2.751	0.0972	***	0.158	0.6912	***
Relative subsistence[a]	1.518	0.2180	***	0.900	0.3428	***	0.828	0.3628	***
Relative subsistence[b]	2.706	0.0999	***	0.995	0.3186	***	6.836	0.0089	Negative
Tenancy	0.349	0.5589	***	0.038	0.8448	***	3.277	0.0703	***
Absolute subsistence	0.043	0.8363	***	0.000	1.0000	***	3.706	0.0542	***
(c) South Rampur (aman HYV rice)									
Own area	1.072	0.3004	***	0.084	0.7716	***	19.718	0.0000	Positive
Operated area	1.072	0.3004	***	0.795	0.3727	***	14.681	0.0001	Positive
Agricultural worker	0.913	0.3393	***	1.540	0.2146	***	1.131	0.2876	***
Labour scarcity	4.274	0.0387	Negative	0.392	0.5315	***	7.154	0.0075	Negative
Absolute subsistence	0.027	0.8694	***	5.416	0.0200	Negative	1.466	0.2260	***
Relative subsistence[a]	9.686	0.0019	Negative	1.005	0.3160	***	19.023	0.0000	Negative
Relative subsistence[b]	8.912	0.0028	Negative	1.554	0.2126	***	15.907	0.0001	Negative
Tenancy	0.372	0.5418	***	0.012	0.9139	***	3.466	0.0626	***
Education	3.604	0.0576	***	1.899	0.1682	***	13.061	0.0003	Negative

*** Not significant at 5 per cent level.

[a] Absolute subsistence/operated area.

[b] Absolute subsistence/own area.

The nature of relationship, e.g. negative or positive, is based on 2 × 2 contingency tables. Chi square values are Yates corrected.

farm. Furthermore, adoption may not take place because of resource constraints as well as higher riskiness during the *kharif* season. As for intensity of adoption, only absolute subsistence has any statistically significant (positive) association with it. The association with other variables is not statistically significant. As expected, index of participation is positively associated with level of education, size of holding, and negatively associated relative subsistence pressure. It seems to be negatively associated with labour supply for much the same reason as mentioned above. Other variables, like tenancy, size of family labour, absolute subsistence pressure, do not seem have any statistically strong association with index of participation.

For Ekdala, crude adoption rate of *boro* HYV is positively associated with irrigation. *Aman* HYV crude adoption rate is positively associated with size of operational holding and irrigated area. Intensity of *boro* HYV adoption is positively associated with irrigation, negatively associated with size of holding and relative subsistence pressure. As for index of participation, it is positively associated with both farm size and irrigated area. All these findings are consistent with *a priori* expectations. Crude adoption rate of *aman* HYV is positively associated with farm size and irrigation. Significant positive association exists between intensity of adoption, per cent area irrigated and educational score. Index of participation is negatively associated with relative subsistence pressure and positively associated with irrigation, education and size of holding.

So far, the hypotheses regarding the effects of various factors have been tested using bivariate technique of analysis. Because of this, the pure effects of different variables are difficult to ascertain and are likely to contain effects of other factors as well. Furthermore, in a 2×2 contingency table (with one degree of freedom) the requirement of any cell with no fewer than 5 expected frequencies could not always be satisfied. As Leabo, (1972, p. 535) suggests, 'even though the expected frequencies are below the requirement, the usefulness of the test may not be destroyed. The results do become inexact though and may cast doubt on the decision'. To investigate the empirical relationships involving various indicators of adoption and other variables, we turn to multivariate analysis in the remainder of this section.

7.5.4 Intensity of Adoption: Multivariate Analysis

In order to investigate empirically possible determinants of the intensity of HYV adoption, we employ least squares regression analysis. The estimated regression equations are set out in Table 7.5. A number of regression equations were estimated but the 'best' ones are reported. These are based on the criteria which the BMDP P9R programme employs to select the 'best' one from all possible subsets of regression (see Dixon, 1983, pp. 264–77). Among all the possible determinants of the intensity of *boro* HYV adoption in Ekdala, farm size (both operational and owned), education and percentage area under irrigation seem to be the most important ones. All the coefficients have expected signs and are highly significant. Th coefficient of the irrigation variable seems to be the most important, followed by farm size and education. Both in terms of explanatory power and statistical significance as indicated by adjusted R^2 and the F-ratio the overall fit can be considered good.

We estimated a similar equation for intensity of adoption of *aman* HYV adoption in Ekdala. Overall fit is not good in terms of explanatory power, even though the F-ratio is significant at the 1 per cent level. All the coefficients are of expected sign and possess statistical significance at the 5 per cent level. The tenancy variable seems to have a negative impact on the intensity of adoption. Furthermore, the percentage area irrigated has also a significantly positive impact on the intensity of *aman* HYV adoption, even though it is primarily a rainfed crop. This is probably because Ekdala, being located in the dry zone, suffers considerably from uncertainty and inadequacy of precipitation. Under such circumstances, farmers may need to provide supplementary irrigation for *aman* HYV cultivation. Those without access to irrigation are unlikely to cultivate *aman* HYV, and even if they do the intensity is unlikely to increase. This is because, as gathered from field observations, in the event of inadequate or untimely rainfall *aman* HYV yields can fall below those of the traditional varieties.

For South Rampur the determinants of intensity of *aman* HYV are different from the ones in Ekdala. Only subsistence pressure and labour supply enter the best possible subset of regression. While the coefficient of the former variable has the expected sign, that of the latter does not. The explanatory power of the equation is poor even though the overall F-ratio is highly significant.

7.5.5 Propensity of HYV Adoption: Logit Analysis

A logit analysis is now used to explain the probability of a farmer adopting HYV. The dependent variable is dichotomous in that it

Table 7.5 Intensity of HYV adoption, Ekdala (Rajshahi) and South Rampur (Comilla), 1985–6: Results of multiple linear regression analysis

Crop	Estimated equation	R^2	F-ratio	DF
(a) Ekdala				
Aman HYV	44.9981 − 8.5630OPERA + 9.6374EDU + 1.3557LABSUP − 8.9672OPERA + 0.2841PCIRRI (2.27)** (2.13)** (1.71)** (2.13)** (1.89)**	0.3309	5.15***	5,37
Aman HYV	47.8197 − 7.7187OWNAREA + 9.5446EDU + 1.6875LABSUP − 11.4115PCOPERA + 0.2555PCIRRI (2.34)** (2.16)** (2.19)** (2.68)*** (1.68)*	0.3367	5.26***	5,37
Boro HYV	8.6920 − 7.7464OPERA + 3.1384EDU + 0.8357PCIRRI (4.54)*** (1.56)* (11.74)***	0.8427	79.59***	3,41
Boro HYV	5.6751 − 7.7632OWNAREA + 3.5012EDU + 0.86693PCIRRI (4.67)*** (1.73)* (12.65)***	0.8457	81.40***	3,41
(b) South Rampur				
Aman HYV	60.0483 − 3.4972ABSUB + 0.4666LABSUP (3.52)*** (1.52)*	0.1883	6.92***	2.49

*** Significant at 1 per cent level.
** Significant at 5 per cent level.
* Significant at 10 per cent level.

assumes a (1,0) value, 1 for adoption and 0 for non-adoption. We do not describe the method in detail here but it can be found in Goldfeld and Quandt (1972, pp.125–34), Theil (1971, pp.628–33) and Kmenta (1971, p.426).[4] We have used the BMDP PLR programme (Dixon, 1983, pp.330–7) to estimate propensity functions. The results are set out in Table 7.6 disaggregated by season and by region.

Initially we included all the variables considered relevant on *a priori* grounds. As expected, propensity to adopt *boro* HYV is positively associated with the size of holding. However, the coefficients of all other variables lack statistical significance. Another set of regression estimates were made using size of irrigated area and other variables. The coefficient of irrigation has the expected sign and is highly significant. Propensity to adopt *boro* HYV is also positively associated with the supply of family labour. During the *aman* season, irrigation and size of holding seem to influence significantly the decision to adopt HYV in Ekdala. We tried with all other variables but none appeared to be significant when included as explanatory variables in the same equation.

For South Rampur the likelihood of *aman* HYV adoption is influenced by relative subsistence pressure. Neither farm size, nor tenancy nor education have any significant association with propensity to adopt. Thus the variables that influence the decision to adopt HYVs in one village are not necessarily the same as those in another. For Ekdala, farm size, irrigated area and relative abundance or scarcity of labour emerge as important determinants of propensity of HYV adoption while relative subsistence pressure seems to be the relevant variable for South Rampur.

7.5.6 Adoption of HYVs: Discriminant Analysis

A prediction study with a nominal rather than a continuous criterion variable calls for a statistical technique known as discriminant function analysis. There are two types of discriminant function analysis: one for dichotomous variables and the other for polychotomous variables. In the present paper the analysis of propensity to adopt involves the use of a *two-group* (e.g. adopter, non-adopter) discriminant function analysis. We do not describe the method in detail here. Based on the discussion of the method in Huck *et al.* (1974, pp.160–5) and Tintner (1965, pp.96–9),[5] we estimate a linear (standardized) discriminant function (see Dixon, 1983, pp.519–35, 683) using variables to discriminate between adopters and non-adopters.

Table 7.6 Propensity of HYV adoption, Ekdala and South Rampur, 1985–6: Results of logit analysis

Innovation	Estimated equation
(a) Ekdala	
Boro HYV	$0.009321 + 1.7384\text{OPERA} - 0.3597\text{PCOPERA} + 0.1009\text{ABSUB} - 0.5893\text{EDU} + 0.1887\text{LABSUP} - 0.0235\text{SUBSIST1}$ $(1.882)**$ $\quad$ (0.7799) $\quad$ (0.4038) $\quad$ (1.041) $\quad$ (1.088) $\quad$ (0.755) Chi-square = 51.834, P-value = 0.441, degrees of freedom = 51
Boro HYV	$-3.1538 + 21.5260\text{IRRI} + 0.2269\text{LABSUP}$ $(3.174)***$ $\quad$ $(1.52)*$ Chi-square = 20.694, P-value = 1.000, degrees of freedom = 55
Aman HYV	$-0.7761 + 8.7704\text{IRRI}$ $(3.025)***$ Chi-square = 42.883, P-value = 0.901, degrees of freedom = 56
Aman HYV	$-0.0294 + 1.4013\text{OPERA}$ $(2.091)**$ Chi-square = 57.197, P-value = 0.430, degrees of freedom = 56
(b) South Rampur	
Aman HYV	$4.1676 - 0.09925\text{SUBSIST1}$ $(2.537)***$ Chi-square = 20.773, P-value = 1.000, degrees of freedom = 56
Aman HYV	$4.1074 - 0.08874\text{SUBSIST}$ $(2.481)***$ Chi-square = 21.921, P-value = 1.000, degrees of freedom = 56

*** Significant at 1 per cent level.
** Significant at 5 per cent level.
* Significant at 10 per cent level.

As with logit analysis, we initially included all the relevant variables in the function. Subsequently, however, variables were selected by the BMDP 7M programme on the basis of the statistical significance of the discriminating variable. The results are set out in Table 7.7 and indicate that for *boro* HYV in Ekdala irrigation has the highest discriminatory power, followed by size of holding (operated or owned). Irrigation also emerges as the only variable with significant discriminatory power for *aman* HYV adoption. It must be noted that the coefficient of farm size does not have the expected sign. A similar result is reported by Asaduzzaman (1979, p.38). In South Rampur, (relative) subsistence is the only discriminatory variable between adopters and non-adopters. However, the minimum D^2 values are low and indicate lower discriminatory power even though the significance of the F-values implies a significant distinction between adopters and non-adopters in either village.

The above analysis has identified factors that affect farm-level adoption of HYV technology in Bangladesh on a static cross-sectional basis. However, one needs to be reminded of the limitations of such an analysis (e.g. Asaduzzaman, 1979; Atiqur Rahman, 1981; R.I. Rahman, 1983). First, the static analysis fails to capture the dynamic behaviour of the diffusion process. The larger farms may be early adopters but it does not necessarily follow that at all times the percentage of adopting farms will be the highest from among the large-farm category. Other conditions remaining the same, the percentage of relatively smaller farms adopting can overtake that of the larger farms. Secondly, static analysis does not allow adequate account to be taken of changes in knowledge and riskiness of adoption with the passage of time. Finally, the welfare significance and distributional implications are not apparent from such an analysis. Let us therefore consider the dynamic pattern of adoption of new technology.

7.6 DYNAMICS OF ADOPTION: OBSERVED PATTERN

To investigate the adoption process in a dynamic context, we proceed firstly to specify the time-pattern of adoption. This is followed by some explanations for the observed pattern of HYV adoption. However, note that the dynamic analysis of adoption is based on answers to the question of initial year of adoption by the respondents. Thus we had to rely heavily on the memory of the respondents, and therefore the data may be subject to some memory bias.

Table 7.7 Propensity of HYV adoption, Ekdala (Rajshahi) and South Rampur (Comilla), 1985–6: Results of Discriminant analysis

Innovation	Standardized discriminant function	F-ratio	DF	Mahalanobis D^2(minimum)	Value at group means Adopter	Non-adopter
(a) Ekdala						
Boro HYV	$-0.7299+4.4894$IRRI$+0.0263$ABSUB $+0.0029$SUBSIST1$+0.2602$PCOPERA -1.1103OPERA-0.2286EDU	2.865	6,51	0.60	0.9356	−0.9355
Boro HYV	$-0.2799+4.3221$IRRI-1.2133OPERA	8.121	2,55	0.10	0.8198	−0.8198
Boro HYV	$-0.2429+4.2228$IRRI-1.1038OWNAREA	10.280	2,55	0.10	1.0377	−1.0377
Aman HYV	$-0.7266+3.0061$IRRI-0.1056ABSUB -0.0062SUBSIST1 $+0.3062$PCOPERA-0.4289OPERA$+0.0488$EDU	2.150	6,51	0.40	0.6369	−0.6372
Aman HYV	$-0.6197+1.7048$IRRI	10.661	1,56	0.10	0.4793	−0.4793
(b) South Rampur						
Aman HYV	$2.9707+0.3153$ABSUB-0.0597OPERA -0.1515SUBSIST1 0.9672PCOPERA-0.0093LABSUP$+0.2609$EDU	7.400	6,51	1.80	4.5314	−4.5314
Aman HYV	$5.6082-0.1542$SUBSIST1	46.062	1,56	0.10	4.2819	4.2814

Table 7.8 sets out data on the time-pattern of (crude) adoption of HYVs by farm size in both South Rampur and Ekdala. In South Rampur, the diffusion of the new technology was complete by the later part of the 1970s when the crude adoption rate for each category of farms reached 100 per cent. In terms of percentage of farms adopting HYVs, the Green Revolution seemed to have been firmly established by the late 1960s. By then more than 75 per cent of farms in South Rampur had adopted HYVs. The picture from Ekdala, however, is quite different. It took a decade longer for the new technology to become widely diffused in Ekdala. Furthermore, whereas South Rampur has reached a crude adoption rate of 100 per cent, Ekdala has not yet been able to do so, primarily because of access to irrigation. If, however, one only considers the adopting farms (those with access to irrigation, i.e. potential adopters) in Ekdala, the diffusion process encompasses a period of 16 years (1969–84) which is no longer than the one for South Rampur (1962–77). This time-lag seems to be consistent with one of more than 14 years required for hybrid seed corn to reach complete adoption in Iowa (Rogers and Shoemaker, 1971, p.16). However, there is a significant difference in the average length of the adoption process between the two villages. Whereas Ekdala took on average a period of more than eight (8.3) years for all (potential) adopters to adopt, for South Rampur the period was much shorter – less than five (4.9) years.

In both villages a higher percentage of larger farmers were early adopters. In South Rampur, subsequently, the percentage of medium and small farms adopting HYVs overtook that of the large farms, even though adoption by all of the latter group was achieved earlier. In Ekdala, the percentage of large farms adopting HYVs was always higher than for smaller farms, not all of whom are adopters. Considering only the adopting farms (those with access to irrigation), the time-pattern changes. 'Half-way' through the adoption process, the percentage of adopting medium farms overtakes that of larger farms, before being overtaken again by the larger farms. These patterns can be more clearly seen in terms of Figures 7.3a–c which plot the data set out in Table 7.8. Thus in general S-shaped adoption curves are observed, increasing slowly at the beginning, at a faster rate thereafter and tapering off towards the end. This seems to be consistent with the patterns reported by Griliches (1957) in his pioneering study on hybrid corn (cf. Ozga, 1960).

Table 7.8   Time-pattern of HYV (all varieties) adoption by farm size in
two Bangladeshi villages: Ekdala (Rajshahi) and South Rampur (Comilla)

(a) Ekdala

Initial year of adoption	Cumulative percentage of farmers adopting HYV						
	Small	Medium	Large	All farms	Small*	Medium*	All farms*
1968	0.00	9.09	14.28	3.45	0.00	10.00	3.92
1969	0.00	9.09	28.57	5.17	0.00	10.00	5.88
1970	7.50	18.18	57.14	15.52	8.82	20.00	17.65
1971	10.00	27.27	85.71	22.41	11.76	30.00	25.49
1972	10.00	27.27	85.71	22.41	11.76	30.00	25.49
1973	10.00	27.27	85.71	22.41	11.76	30.00	25.49
1974	10.00	27.27	85.71	22.41	11.76	30.00	25.49
1975	20.00	27.27	85.71	29.31	23.53	30.00	33.33
1976	30.00	36.36	85.71	37.93	35.29	40.00	43.14
1977	50.00	63.64	85.71	56.90	58.82	70.01	64.71
1978	52.50	63.64	85.71	58.62	61.76	70.01	66.67
1979	62.50	81.82	85.71	68.97	73.53	90.01	78.44
1980	72.50	81.82	85.71	75.86	85.29	90.01	86.27
1981	72.50	81.82	100.00	77.59	85.29	90.01	88.24
1982	72.50	81.82	100.00	77.59	85.29	90.01	88.24
1983	80.00	90.90	100.00	84.48	94.12	100.00	96.08
1984	85.00	90.90	100.00	87.93	100.00	100.00	100.00

(b) South Rampur

Initial year of adoption	Cumulative percentage of farmers adopting HYV			
	Small	Medium	Large	All farms
1962	5.71	13.33	25.00	10.34
1963	17.14	13.33	25.00	17.24
1964	28.57	20.00	50.00	29.31
1965	60.00	66.66	50.00	60.34
1966	71.43	73.33	62.50	70.69
1967	74.29	86.67	75.00	77.59
1968	74.29	86.67	75.00	77.59
1969	74.29	86.67	75.00	77.59
1970	74.29	86.67	75.00	77.59
1971	74.29	86.67	75.00	77.59
1972	77.14	86.67	75.00	79.31
1973	80.00	86.67	75.00	81.03
1974	85.71	86.67	87.50	86.21
1975	88.57	93.33	100.00	91.38
1976	97.14	93.33	100.00	96.55
1977	100.00	93.33	100.00	98.28
1978	100.00	100.00	100.00	100.00

Note: Columns marked with asterisks are derived on the basis of the number of total
adopting farmers in each category. The Ekdala data also include those who adopted
but subsequently dropped out.

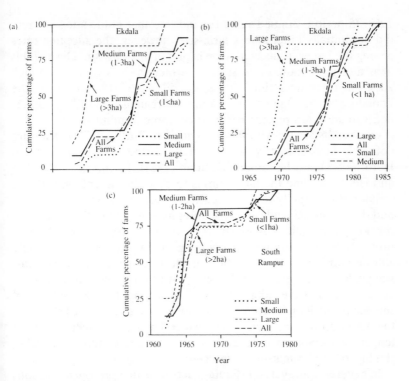

Figure 7.3   Time pattern of adoption of HYVs of: (*a*)   All farms in each category, Ekdala (Rajshahi); (*b*)   Adopting farms in each category, Ekdala (Rajshahi); (*c*)   All farms in each category, South Rampur (Comilla)

How do the adoption patterns in our sample villages compare with those reported in recent studies? Hayami and Ruttan (1984, p.50) report findings of an International Rice Researh Institute study (IRRI, 1978) on the adoption of HYVs in 30 Asian villages during the 1966–72 period. Their study indicates that small and medium farms were earlier adopters, even though large farms did not lag far behind. Furthermore, during the adoption process the medium farms overtook the smaller farms who were initially the faster adopters. However, one must take note of the highly aggregative nature of the data concealing the diversity of underlying environmental, ecological and socio-economic factors. I. Ahmed (1981), using surveys in three Bangladeshi villages, reports the dynamic adoption pattern for the

1968–75 period. Ahmed's findings indicate that crude adoption rate did not initially differ across farm-size; however, the adoption rates among the larger farms tend to be higher in the long run. Despite this 'it is also to be noted that smaller farmers have also attained high adoption rates over time' (pp.25–6 and Table 10). However, the limitation of the aggregative nature concealing inter-village differences in adoption pattern needs to be borne in mind.

Our results accord with the contention of Ruttan (1977). They are definitely inconsistent with those of I. Ahmed (1981), Barker and Herdt (1978) and Hayami and Ruttan (1984). In contrast to Barker and Herdt, we find the larger farms to be earlier adopters, with the smaller farms rapidly catching up. In contrast to I. Ahmed (1981), we find the gap in the crude adoption rates narrows rather than widens in the long run, and cross-overs even occur at some stage. This is probably because Ahmed covers the period up to 1975 which could still be considered the inception of the Green Revolution in most Bangladeshi villages. Even though most of the larger farms in Bangladesh might have adopted HYVs by the mid-1970s, holders of smaller farms had yet to overcome their psychological resistance to the adoption of a new and unknown technological package. Had a longer time-span been considered, a different picture (somewhat similar to ours) might have emerged.

In general, however, our findings accord with *a priori* expectations that larger farms are earlier adopters, even though smaller farms may not lag far behind them. Bangladeshi empirical evidence also supports our findings.

7.7 EXPLAINING THE OBSERVED TIME-PATTERN OF ADOPTION

Why does the adoption pattern behave the way it does in the two sample villages? What factors explain inter-village and intra-village differences in the pattern of HYV adoption?

Differences in the timing of exposure to information on HYVs and installation of modern irrigation equipment (low lift pumps (LLP); shallow tubewells (STW) and deep tubewells (DTW)) may account for inter-village differences in the adoption of new technology. In South Rampur the first of the modern irrigation machinery, an STW, was installed in 1962, and the installation of a few other LLPs, STWs and DTWs followed in quick succession. South Rampur, being located in the laboratory area of the Bangladesh Academy for Rural Development (BARD), was one of the earliest to experience the introduction of the modern variety rice, *Paijam*,[6] in 1962. In Ekdala, on the other hand, the first modern irrigation equipment, a DTW, was sunk in 1969, which was accompanied by the introduction of HYVs of rice.

The difference in average length of the diffusion period processes may be attributable to the speed with which complementary inputs, e.g. irrigation, have spread or become available. The initial years of the BARD were characterized by supply of subsidized inputs, e.g. chemical fertilizers, irrigation and pesticides and access to credit. These raised (private) profitability of innovations and, along with a developed information base (through BARD), reduced risk for all classes of farms. This may have been the catalyst for the rapid diffusion of the new technology among most farms in South Rampur. For Ekdala, however, these factors were less favourable. There was no rapid spread of irrigation. By the time the Green Revolution in Ekdala was beginning to set in, these subsidies were drastically reduced, which curtailed private profitability of innovations. This is one of the reasons for the longer average period for the diffusion process of Ekdala.

Interaction between the degree of uncertainty associated with the innovation and the prevalence of risk aversion among cultivators significantly determines their adoption behaviour. Risk aversion is an endogenous factor. Change in farmers' perceptions may lead to changes in the implications of risk aversion in terms of decision-making. As the new and 'untried' technology is perceived as more risky because it is largely unknown, and assuming that small farms are more risk averse (see, for example, Shahabuddin *et al.*, 1986; Feder and O'Mara, 1981), it may be argued that smaller farms are less likely to adopt it initially.

Traditional technology gives a lower expected return with a greater degree of certainty, and does not require any specialized inputs. HYVs on the other hand realize their potential only if adequate and timely application of fertilizers, irrigation water and pesticides occurs. A further requirement is improved agronomic practices, e.g. proper timing of planting and weeding. But the expected return per hectare of land planted with HYVs is fraught with considerable uncertainty determined by objective and subjective factors. As for the former, HYVs are more susceptible to diseases, pest attacks and climatic, environmental and ecological variations (Tisdell, 1983a).

Schutjer and Van der Veen (1977, p.23) argue that 'the new wheat varieties introduced in the mid 1960s probably increased the uncertainty facing farmers by. . .introducing genetic homogeneity in the variety planted' (quoted in Feder, 1980, p.264; see also Biggs and Clay, 1981, p.330; Feder, 1982). Furthermore, dependence on timely availability of specialized inputs introduces another element of uncertainty. Subjective uncertainty is an inevitable concomitant of any new technology. As Lipton (1978, p.323) emphasized,

> risk is in the eye of the beholder. . . .HYV research successfully stresses robustness. . . .Indeed many areas may feature HYVs with not only higher expected net return, but also a lower risk of disastrously low return than competing traditional varieties. However, for the farmer is familiar with only traditional varieties, it is the HYVs' subjective risk that counts.

One would expect, however, that with the passage of time, information about an innovation would become more easily available and reliable. Moreover, the experience gained by early adopters is likely to affect other farmers' perception of innovations. Also research and development (R & D) may eliminate some uncertainty.

To what extent does the above help explain the observed time-pattern of HYV adoption in the two Bangladeshi villages? This is closely related to the factors that account for the inter-village difference in the average lengths of the diffusion process. South Rampur being an experimental village of the BARD, dissemination of *information* on new technology was quite rapid. However, the difference in time-pattern for different categories of farms may have occurred partly because of the time-pattern of the availability of controlled irrigation. This might have been particularly true in case of Ekdala. Our field survey indicated that the first DTW was installed on large farms and was accessible in the main to the larger farms. Even if all farmers had equal access to irrigation at the same time, some farms might have adopted earlier than others, partly because of differences in subjective risk, and partly owing to variations in absolute and relative subsistence pressure. Furthermore, perception of profitability of the innovation might have been different for different categories of farms. Apart from this, reluctant acceptance of innovations by smaller farms at the beginning and their rapid adoption subsequently may be due to two factors.

First, with their weaker risk-bearing capacity, the majority of the smaller farmers wait till they are almost certain of the demonstrated superiority of new technology over the traditional one. Thus when some of the farms (especially smaller farms) have planted HYVs and these are found to be successful, the others adopt at a rapid rate. This is akin to Mansfield's (1961, p.762) hypothesis that the probability that a firm will introduce a new technique is an increasing function of proportion of firms already using it. This emphasizes the role of imitation as a factor in diffusion of innovations.

Secondly, qualitative improvements in the innovations (HYVs) themselves took place over time. Farmers tend to replace earlier HYVs by later varieties more adapted to their climatic and ecological conditions. Some evidence of the above phenomenon was found from the two survey villages. In South Rampur, *Paijam*, Taipei, IR-5 and IR-8 were the main rice varieties during the initial stages of adoption. They have been gradually replaced by the varieties developed by the Bangladesh Rice Research Institute (BRRI). Prominent among those are BR-3 and BR-11. In the survey year (1985–6) farmers were found to allocate over 70 per cent of all HYV rice area to four rice varieties: BR-3 (21 per cent), BR-11 (26 per cent), *Paijam* (13 per cent) and Taipei (10 per cent). In Ekdala, on the other hand, the varieties that were adopted in the initial stages are IR-8, China and BR-11. At present, these last two varieties dominate the HYV rice hectarage in that during the 1985–6 season these two varieties constituted 82 per cent of the gross area planted to HYV rice. IR-8, which spearheaded the Bangladeshi Green Revolution in the late 1960s, is planted only in 10 per cent of the HYV rice area in Ekdala.[7]

In order to identify the factors that underlie the intra-village time-pattern of adoption, we have differentiated between 'innovators' and 'imitators'. Based on time-distribution of adoption, Shetty (1968, p.1277) divides the farmers into three groups: *(a)* innovators (whose adoption period preceded the average year by one standard deviation); *(b)* imitators (whose adoption followed later); and *(c)* non-adopters.[8]

The period required for adoption of HYVs by farms in Ekdala has a mean value of 8.3 years with a standard deviation of 4.4 years. The corresponding figures for South Rampur are 4.9 years and 4.6 years respectively. While the Shetty criterion gives an innovator–imitator dividing line of 4 years for Ekdala and 0.3 years for South Rampur, the latter is an over-short dividing line.

For the purpose of this study, we have used half the average

Table 7.9   Selected socio-economic characteristics of innovators, imitators and non-adopters in two Bangladeshi villages: Ekdala (Rajshahi) and South Rampur (Comilla)

Characteristics	Average values				
	Ekdala			South Rampur	
	INNOV	IMIT	NADOPT	INNOV	IMIT
Own area (ha)	2.487[a]	0.853	0.457	1.075	0.800
Operated area (ha)	2.486[b]	0.858	0.463	0.994	0.862
Family size (adult equivalent)	7.700	5.026	4.771	4.794	4.727
Subsistence pressure (family size/ operated ha)	3.100[c]	5.860	10.440	4.820	4.480
Subsistence pressure (family size/owned ha)	3.100	5.890	10.300	4.460	5.910
Educational score	2.019[c]	1.030	1.071	1.147	1.159
Irrigated area (ha)	1.022[a]	0.412	0.000	0.994	0.862
Area rented-in (% of own area)	3.660	18.700	20.790	6.605	10.630
Information base: Informal only (% of group total)	39.47	38.46	NA	00.00	00.00
Both sources (% of group total)	57.89	61.54	NA	80.48*	23.53
Number of farmers	38	13	7	41	17

Note: INNOV, IMIT and NADOPT respectively refer to innovators, imitators and non-adopters; [a], [b] and [c] respectively refer to 1, 5 and 10 per cent level of significance for chi-square values; * means significant at 1 per cent level for test of difference of proportions. There were no non-adopters in South Rampur.

number of years acquired for the adoption by farms as the cut-off point. According to this criterion we have an unchanged dividing line of approximately 4 years for Ekdala and 2.5 years for South Rampur. On the basis of this classification we discuss below the socio-economic characteristics of innovators, imitators and non-adopters. The results are set out in Table 7.9. We found 38 imitators, 13 innovators (and 7 non-adopters) in Ekdala, while South Rampur had 41 imitators and 17 innovators.

The information contained in Table 7.9 clearly indicates that the Ekdala data support the hypothesis that holders of larger farms with better access to irrigation, education, and a larger information base are the innovators. In the South Rampur area the importance of farm size as an influence on the propensity to innovate does not show up, probably because of close proximity of the area to the BARD.

Let us now specifically investigate the extent to which sources of information differentiate between an innovator and an imitator. We distinguish between two sources of information, namely, 'formal' and 'informal'. Sources such as direct contact with extension agents, membership of farmers' cooperatives, are designated as 'formal' channels, while other sources, e.g. 'seeing others' (i.e. neighbours and friends) as 'informal'. Fifteen out of 38 imitators and 5 out of 13 innovators in Ekdala indicated their knowledge of new technology from informal sources only. None of the innovators seemed to have depended exclusively on formal sources, while only one imitator was reported to have knowledge of the new technology from only formal sources. Twenty-two of the imitators and 8 of the innovators in Ekdala depended on both sources of information. The picture from South Rampur, however, is entirely different. There, none depended entirely on informal sources. Among the 41 imitators, 33 depended on both types of sources. On the other hand, only 4 of the 17 innovators depended on the combination of the two sources. Thus in South Rampur innovators depended more on formal sources, while the opposite was true for imitators.

In order to see the statistical significance of the differences in percentage of adopting farms depending on two types of information sources, we carried out tests of differences of proportions (see Hamburg, 1970, pp.338–41). Our test clearly indicated that there was no evidence to suggest that the Ekdala innovators were less dependent on informal sources than imitators. Similar findings emerged when the test was carried out for farms in both categories reliant on a combination of formal and informal sources. For South Rampur, however, the difference between percentage of innovators depending on formal sources was highly significant.

In both villages informal sources play a significant part in the diffusion of new technology. In Ekdala, even though the majority of the farmers have contacts with the extension agents, adoption by 'seeing others' seems to be a critical factor in the spread of the new technology. Even the innovative farmers' decisions to adopt seem to depend on observing demonstrated success in neighbouring villages.[9] In South Rampur while a small percentage of the innovators depend on indirect sources of information, a vast majority of the imitators seem to base their adoption decision on seeing success of others with the new technology.

7.8   DYNAMICS OF INTENSITY OF ADOPTION

So far we have considered crude adoption rates and in particular the relationship between farm size and the relative frequency of adoption of HYVs. Let us now consider the relationship between intensity of adoption of HYVs (and related innovations) and size of farms, using our available data. Our evidence indicates that the intensity of adoption of HYVs and use of related innovations tends to increase with the size of farms. This aspect has not received much attention in previous studies investigating dynamics of adoption (e.g. Herdt, 1987; Barker and Herdt, 1978; Schluter, 1971). This pattern has some implications for distribution of higher output gains from adoption of new technology, as for instance observed by Feder *et al.* (1985, p. 288).

We do not have data over a long period of time. However, our data for a six-year period (1980–1 to 1985–6) from two survey villages may be indicative of an emerging pattern. Table 7.10 presents relevant data for various innovations. The following pattern is indicated:

1. In Ekdala, the area of small farms under HYVs and the area under irrigation show a declining tendency over time whereas, for medium and large farms, irrigated area and HYV areas show some increase and the total area under HYVs (wheat excepted) remain much the same over time.
2. In South Rampur, the area under *aman* HYV shows an increasing tendency over time for all classes of farms. The individual areas under *boro* HYV remain constant as does their total, since the entire cultivable land in the village is irrigated and is under HYV during the *rabi* season.
3. In Ekdala, medium-sized farms use the highest amount of fertilizers per net cropped hectare. In South Rampur, the large farms use the highest amount of fertilizers per hectare, followed closely by medium and small farms. Also note the striking difference in fertilizer usage by farm size in Ekdala and South Rampur.
4. In both the villages, the use of chemical fertilizers has increased quite rapidly in recent years. It is more rapid in the case of Ekdala. Furthermore, all classes of farms use considerably higher amounts of fertilizer in Ekdala than those in South Rampur.

Let us now examine what implications the observed pattern of adoption might have on the distribution of innovations. To do this,

Table 7.10 Trends in area under different HYV foodgrains and use of related innovations by size of farms: Ekdala (Rajshahi) and South Rampur (Comilla), 1980–1 to 1985–6

(a) Total area (hectares) under HYV crops by farm size

(i) Ekdala

Year	Aman rice				Boro rice				Wheat			
	Small	Medium	Large	Total	Small	Medium	Large	Total	Small	Medium	Large	Total
1980–81	10.49	7.48	7.28	25.25	11.55	5.94	4.81	22.30	1.17	2.20	3.07	6.44
1981–82	10.29	7.81	8.28	26.38	11.82	6.21	5.28	23.31	1.30	2.60	4.54	8.44
1982–83	8.50	8.15	8.75	25.40	9.91	6.28	5.28	21.47	1.77	2.47	5.41	9.65
1983–84	6.84	8.15	8.75	23.74	7.85	6.41	6.34	20.60	1.80	3.47	5.94	11.21
1984–85	6.77	8.39	9.15	24.31	8.27	6.94	5.68	20.89	1.60	3.07	6.48	11.15
1985–86	7.01	8.12	10.95	26.08	8.36	6.48	4.81	19.65	1.30	2.93	7.75	11.98

(ii) South Rampur

Year	Aman rice				Boro rice			
	Small	Medium	Large	Total	Small	Medium	Large	Total
1980–81	4.65	5.10	5.99	15.74	14.68	18.23	18.45	51.36
1981–82	4.82	5.02	5.91	15.74	14.68	18.23	18.37	51.28
1982–83	5.52	5.83	6.92	18.27	15.09	18.15	18.37	51.61
1983–84	6.13	6.15	5.92	18.20	14.68	17.26	18.37	50.31
1984–85	6.45	6.88	6.80	20.13	15.70	17.30	16.27	49.27
1985–86	6.72	7.85	6.39	20.96	16.34	18.85	16.84	52.03

continued on page 154

Table 7.10 continued

(b) Area under irrigation (ha)

	(i) Ekdala				(ii) South Rampur			
Year	Small	Medium	Large	All	Small	Medium	Large	All
1980–81	12.05	9.28	11.55	32.88	14.33	18.23	18.45	51.01
1981–82	12.38	9.22	11.95	33.55	14.33	18.23	18.45	51.01
1982–83	11.48	9.35	12.35	33.18	14.73	18.15	18.45	51.33
1983–84	9.19	9.35	12.15	30.69	14.89	17.58	18.45	50.93
1984–85	9.68	9.75	12.69	32.12	16.27	17.09	16.35	49.71
1985–86	9.12	9.82	11.49	30.43	16.34	13.85	16.84	52.03

(c) Fertilizer used per net cropped ha (kg)[a]

	Small	Medium	Large	All	Small	Medium	Large	All
1980–81	503	756	690	591	283	302	339	296
1981–82	549	784	646	608	283	305	339	297
1982–83	602	901	711	670	314	346	360	329
1983–84	647	943	748	713	379	411	439	396
1984–85	658	954	759	724	402	458	468	425
1985–86	769	1091	935	849	536	579	608	557

[a] Measured in terms of gross weight of fertilizers as averages of those households who actually used chemical fertilizers.

Table 7.11   Indices of concentration of HYVs and related innovations in two Bangladeshi Villages: Ekdala (Rajshahi) and South Rampur (Comilla), 1980–1 to 1985–6

Innovation	Indices of concentration					
	1980–1	*1981–2*	*1982–3*	*1983–4*	*1984–5*	*1985–6*
(a) Ekdala						
Aman HYV	0.293	0.323	0.382	0.430	0.441	0.463
Boro HYV	0.180	0.193	0.239	0.329	0.305	0.271
Wheat HYV	0.557	0.600	0.582	0.592	0.623	0.674
Rabi HYV	0.265	0.302	0.346	0.422	0.416	0.424
All HYV	0.278	0.311	0.362	0.425	0.427	0.441
Irrigated area	0.356	0.355	0.380	0.429	0.427	0.423
(b) South Rampur						
Aman HYV	0.362	0.351	0.355	0.304	0.323	0.310
Boro HYV	0.362	0.361	0.355	0.359	0.322	0.323
All HYV	0.362	0.350	0.355	0.344	0.322	0.319
Irrigated area	0.367	0.367	0.361	0.357	0.314	0.323

Note: *Rabi* HYV includes HYVs of wheat and *boro* rice. The indices of concentration relate to all sample farm households.

we have calculated an index of concentration somewhat similar to the Gini ratio for various HYVs and irrigated area in the two villages for the 1980–1 to 1985–6 period. The Gini concentration ratio involves the use of the same variable on both axes. For instance, if one derives the Gini coefficient of land ownership distribution, the horizontal axis measures x per cent of owners while the vertical axis measures the corresponding (say y per cent) farm area (see Yotopoulos and Nugent, 1976, p.242). The index of concentration of various elements of the new technology used in this paper differs from the above. Here the two axes measure two different variables. For instance, we have considered x per cent of landowners (along the horizontal axis) having control over y per cent of irrigated area (along the vertical axis).

The results are presented for all farms in our sample in Table 7.11 and are illustrated for some innovations (for the Ekdala sample) in Figure 7.4. The results show a contrasting pattern of distribution of the new agricultural technology for the two villages. For Ekdala there is an increasing tendency towards concentration for all innovations. Between 1980–1 and 1985–6 concentration ratios for *aman* HYV, *rabi* HYVs, all HYVs and irrigated area increased respectively by 58,

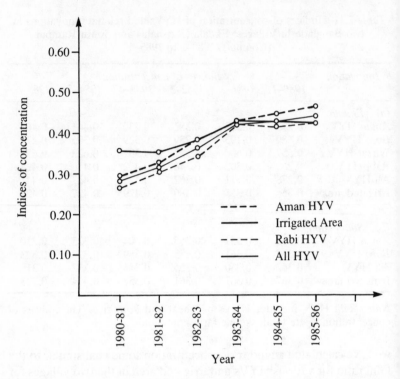

Figure 7.4   Trends in indices of concentration for adoption of HYVs and irrigation, Ekdala (Rajshahi), 1980–1 to 1985–6

62, 58 and 19 per cent. This pattern seems to be consistent with that for Bangladesh as a whole (see also Chapter 8). For South Rampur on the other hand, the concentration indices remain much the same with a weekly declining tendency.

The differential behavioural pattern of concentration ratios may be the result of two factors. First, even though most Ekdala farms have adopted HYVs, the intensity of adoption is far below that of South Rampur where a more stable pattern seems to have been in existence for some time. Secondly, and perhaps more importantly, South Rampur has a more equitable (Gini ratio = 0.415) land ownership distribution pattern, whereas in Ekdala inequality in land distribution is quite high (Gini ratio = 0.553). Transmission of further income inequality through an increasing concentration of innovations cannot be ruled out (Hayami and Ruttan, 1984, p.49).

Thus although the intensity of adoption of HYVs and related

technology tends to be higher for small farms than the large ones on those farms adopting HYV at a point of time (Alauddin and Tisdell, 1988e) as for instance found by Muthia (1971), Schluter (1971) and Sharma (1973) in India and by I. Ahmed (1981) and Asaduzzaman (1979) for Bangladesh, we do not find evidence in Bangladesh to support Schluter's contention for India that the degree of this relationship increases with the length of time since the introduction of new varieties. Indeed an opposite trend is apparent in Ekdala, and in South Rampur the concentration ratios seem relatively stable. Indeed it is possible that after the adoption of HYV innovation, concentration ratios tend at first to increase and eventually come to relatively stationary (equilibrium) values. So at least initially there appears to be some increase in dualism (cf. Yotopoulos and Nugent, 1976, p.238). However, more research is needed and we must not generalize from our sample. Nevertheless, it is clear that Schluter's contention cannot be generalized.

7.9 CONCLUDING OBSERVATIONS

Employing bivariate and multivariate techniques of analysis, and utilizing primary data from two different villages, it has been found that degree of access to irrigation emerges as the key determinant of HYV adoption both within and between the villages. All the indicators of adoption, *crude adoption rate, intensity of adoption, index of participation* and *propensity to adopt* are significantly influenced by the irrigation variable. Other important determinants are farm size, labour scarcity and relative subsistence pressure.

Significant differences between villages exist in the adoption of technology during the dry season. There is 100 per cent *crude adoption rate* of HYVs as well as 100 per cent *intensity of adoption* by every farm in South Rampur during the dry season, which contrasts with the picture at Ekdala. However, little difference exists in both the measures of adoption of HYVs during the rainy season. It needs to be pointed out that in Ekdala primarily those farmers who have access to irrigation adopt HYVs in the rainy season, so that in case of inadequate and uncertain rainfall supplementary irrigation can be arranged. Significant differences between villages exist in the adoption of technology during the dry season. In South Rampur, since everyone irrigates, the significance of irrigation does not show up from statistical analysis. Less than 100 per cent *crude adoption rate* and a little more than 40 per cent *intensity of adoption* of rainy season HYVs in South Rampur is due to the fact that present HYVs are insufficiently flood-resistant. Adoption of HYVs under flood-prone conditions is a more risky proposition than under assured source of irrigation during the dry season. Nor are they very drought-resistant, as is evident from their adoption in the drought-prone village of Ekdala. In case of a severe flood or a drought the yields of these varieties may fall below those of traditional varieties. Under the present circumstances, it is unlikely that for the rainy season the area under HYV will increase much further (see also Bera and Kelley, 1990). The technological and environmental constraints apart, 'there are long-established tastes for certain types and kinds of grains which are not satisfied by the present HYVs. . . .the HYVs are unlikely to completely replace traditional varieties. . .in the near future' (Dalrymple, 1976, p.71).

On the basis of *all* farms in the samples, there is support for the hypothesis that bigger farms are earlier adopters but the smaller farms rapidly catch up with larger farms. Initially in both of our sample villages a higher percentage of larger farms were early adopters than were smaller farms. Subsequently, in South Rampur, the percentage of medium and small farms adopting HYVs overtook that of the larger farms, even though adoption by all of the latter group was achieved earlier. In Ekdala, the percentage of large farms adopting HYVs was at all times higher than for smaller farms, and not all of the small and medium farms have adopted HYVs. If, however, one considers *only* adopting farms (those with irrigation), at some stage during the diffusion process the percentage of medium-sized farms adopting overtakes that of the larger farms, before being overtaken again by larger farms. This pattern is evident both in Ekdala and in South Rampur.

The analysis of the characteristics of innovators and imitators provides additional support for the above hypothesis from the Ekdala data, while no firm conclusion can be drawn from the South Rampur data. The involvement and proximity of the South Rampur village to official crop experiments, trends in subsidization of inputs and frequent visits by the members of the BARD may have made it atypical of the Bangladeshi situation.

Even though innovators and imitators alike have contacts with extension agents or have access to formal channels of information, it is the demonstrated success (e.g. seeing others' success: i.e., informal channels of information), that seems to have had the greatest impact on decisions to adopt HYV. Even the innovative farmers of Ekdala seem to have been influenced by seeing the success of new technology in neighbouring villages.

The dynamics of adoption has implications for income distribution. There is evidence of increased concentration by farm size in the usage of vital components of the new technology package. One further complication is that farms report increased fertilizer requirements to maintain yield (see, for example, Alauddin and Tisdell, 1987; Jones, 1984; R.I. Rahman, 1983). If this is the likely trend, smaller farms may find it more difficult to maintain yield *vis-à-vis* that of larger farms. This could further accentuate inequality. Thus the evidence of eventual equalization of adoption rates of HYV technology by various farm categories may not in itself be an insurance against increased inequality in the distribution of income and ownership of assets. Our findings seem to support the Feder *et al.* observation that

> even if this is the case, the early adopters (usually the larger and wealthier farms) can accumulate more wealth and use the differential in the subjective value of land to acquire more land from the laggards. The accumulation of new wealth enables further adoption and thus affects the dynamic pattern of aggregate adoption (Feder *et al.*, 1985, p.288).

While our case study is based on only two villages, and one therefore needs to exercise caution in generalizing from it, it does provide evidence in favour of some hypotheses in the literature and casts doubt on others about nature of the dynamics of adoption and diffusion of HYVs by farms of different sizes. Furthermore, there are indications that some of the observed patterns may generalize to the Bangladeshi scenario as a whole, e.g. the increasing relative concentration of use of vital inputs in the HYV technology package in the hands of those holding larger-sized farms.

8 Poverty, Resource Distribution and Security: The Impact of New Agricultural Technology in Rural Bangladesh

8.1 INTRODUCTION

Increasing inequality in use of land and of access to natural resources essential for the success of modern agricultural techniques can be expected to increase inequality in rural income distribution, especially if the number of landless or near-landless rises and their real per capita income either falls or remains constant. This chapter provides evidence of increasing concentration of use and control of agricultural land in Bangladesh, and of increasing inequality in the ownership of ancillary resources such as irrigation water, essential for the success of the bulk of Green Revolution technologies. Increasing landlessness and near-landlessness is making the Bangladeshi rural poor more dependent on wage employment for their subsistence. How the poor fare in these circumstances depends in the main on the availability of wage employment and the real level of wages. In this regard, A.R. Khan (1984, p.192) found that despite some short-term recovery due to rapid output growth, the long-term trend in agricultural wages is negative. Slow growth in agricultural productivity is one of the major factors explaining the negative trend in real wages. Khan identified (p.194–5) four long-term factors which tended to depress real agricultural wages in Bangladesh during the 1949–82 period: (1) declining land–man ratio; (2) continued inequality in the distribution of land; (3) the prevalence of institutions and techniques which discouraged increased on-farm employment; and (4) slow growth in off-farm employment. In an earlier study, R. Islam (1979) found evidence of increased poverty and inequality in terms of falling

160

real wages and increased concentration of land resources.

However, the actual increase in the incidence of rural poverty and the degree of inequality may be much greater than is apparent from such studies which are based only on the analysis of exchange income and ignore non-exchange income. Note that two different approaches have been employed in analysing the impact of agrarian technological change on income and resource distribution and rural poverty: (1) analysis based on exchange income (see, for example, Ahluwalia, 1978; 1985; Mellor, 1978; 1985; Hayami and Herdt, 1977); and (2) focusing primarily on non-exchange income (e.g. Jodha, 1985; 1986; Lipton, 1985; Conway, 1985).

The income of the rural poor (the landless and the near-landless) consists of two components: exchange income (primarily wage income) and non-exchange income and, in our view, both components are important. The first is determined largely by market forces. The second is determined primarily by institutional and sociological systems in the rural society and is usually obtained directly from nature without exchange, and includes, among other things, wild fruits, wild animals, firewood, building and thatching materials, water from tanks, streams and ponds for growing vegetables and fruit mainly for domestic consumption, free-ranging of poultry and some limited grazing or fodder for livestock, e.g. sheep, goats and cattle. These sources of income depend to some extent on common access or low-cost access to natural resources.

Resources available as a source of non-exchange income appear to be becoming scarcer in many LDCs (Jodha, 1985; 1986; Conway, 1985). Population pressure has depleted them and greater penetration of technological and market forces many have led to some resources being priced or access to some of them being restricted. Again the traditional 'sharing ethics' at the local and community levels may have been or are gradually being replaced by egocentric considerations characteristic of more individualistic societies (Tisdell, 1983a). These developments may have adverse welfare implications for the rural poor, consisting of the landless and the near-landless, since their overall welfare is determined by the relative importance of the exchange and non-exchange components of their overall income.

While studies on rural poverty in Bangladesh are substantial, they focus primarily on the exchange component of incomes of the rural poor and neglect the non-exchange component. In our view, important aspects (e.g. distribution of ancillary production resources, access to common property resource and other sources of non-exchange

income) having implications for rural poverty have not been discussed for Bangladesh. We intend to give particular attention to
these in this chapter.

We proceed first of all by analysing the distribution and ownership
of land and other resources using Bangladeshi agricultural census
data as well as farm-level data. Where possible, trends in the concentration and control of resources are highlighted. This is followed
by an analysis of the sources of market and non-market income of
Bangladeshi rural poor, initially by presenting *a priori* considerations
and subsequently by providing empirical evidence from Bangladesh.

8.2   LAND-USE CONCENTRATION, LANDLESSNESS AND DISTRIBUTION OF ANCILLARY RESOURCES: ANALYSIS OF AGRICULTURAL CENSUS DATA

With rapid population growth, Bangladesh has experienced growing
landlessness in recent years. The percentage of rural landless households is on the increase, implying growing inequality in the distribution of land resources. However, considerable conceptual difficulties
and statistical pitfalls surround the land statistics which are available
for the years 1960, 1968, 1977 and 1983–84 (see, for example,
Abdullah, 1976, p.90; Abdullah *et al.*, 1976, pp.212–3; Cain, 1983,
pp.154, 158). The data for 1960, 1968 and 1977 are based on sample
surveys and involve only a small fraction of rural households,
whereas the 1983–4 data are based on complete enumeration. Furthermore, the data relating to 1960 and 1968 do not provide information on a land ownership basis. The 1977 and 1983–4 figures provide
data on land distribution on both an operational and ownership basis.
In addition, differences in definitions and scope of various census
data pose difficulties for inter-temporal comparisons of land distribution.[1]

Analysis of data from various agricultural censuses (ACO, 1962,
p.29; BBS, 1972, p.9; 1981, p.41; 1986b, p.81) indicates a substantial
degree of inequality in land distribution. This is reflected in very high
values of the Gini concentration ratios. If one compares the Gini
concentration ratio (0.502) for operational land from the 1960 Census
(ACO, 1962) with that (0.479) of the 1967–8 Master Survey of
Agriculture (BBS, 1972), it might seem that there has been a trend
towards equalization of land holdings. However, this is primarily a
matter of concentration of holdings in the small-farm category (less

than 0.40 ha). The fall in the value of Gini coefficient (0.419 for operational holdings) in 1977 is due probably not to a trend towards greater equalization but an under-reporting of the number of small farm households (BBS, 1986b, pp.32–3). If one considers the pattern of land ownership distribution for 1977 and 1983–4 an interesting feature emerges: the 1977 Gini coefficient is estimated to be 0.428, which is higher than the one (0.419) for operational holdings in that year. Thus the operational holdings seem to be slightly more evenly distributed than owner holdings. The 1983–4 data, however, indicate that land distribution on ownership basis is more evenly distributed than operational holdings (the ownership Gini coefficient is 0.493, as against the one of 0.533 for operational holdings). This suggests the emergence of sharecropping of a 'reverse' pattern whereby small farmers as a group are net lessors while relatively larger farmers are net lessees. This may be because of:

1. Higher profitability of self-cultivation by larger farmers (Bardhan, 1984, p.189; Quasem, 1986, p.18);
2. Greater control by medium and larger farmers over (*khas*) land with or without any formal ownership rights; and
3. Ownership of irrigation equipment by larger farmers giving them almost absolute control over irrigation water. The small farmer dependent upon irrigation supplies from large farmers may sometimes be forced to rent out land to larger farmers who, with relatively easier access to irrigation water and other complementary inputs, are able to appropriate greater relative gains than the traditional sharecroppers (e.g. small farmers). Even when the ownership of irrigation is cooperative, the smallholder usually becomes very much an unequal partner. The siting of tubewells normally takes place on the large farmer's plot, giving him control and easier access to irrigation water (R. Islam, 1979). This phenomenon seems to be supported by our field observation from Ekdala. Furthermore, through the patron–client relationship, the larger farmers who own irrigation equipment gain substantial revenues as rents for irrigation water sold to smaller farmers. The whole process seems to have resulted in the creation of a class generally known as 'water lords' in the rural society (Alauddin and Tisdell, 1986d; Boyce, 1987b).

Thus access of the rural poor to land resources is gradually becoming increasingly limited.

The census figures for 1960 do not provide estimates of the percentage of rural households owning no lands. Cain (1983, p.158), employing some assumptions, reports 17 per cent of the rural households to be landless in 1960. Employing the 1961 Population Census data, Abdullah *et al.* (1976, pp.212–13) derived an estimate of the percentage of households with no land to be 21.9. Subsequent surveys and census reports in the 1970s indicated significant increases in landlessness. For instance, an IRDP (Integrated Rural Development Programme) benchmark survey, based on a subsample of 14 villages and a very small number of households, reported a figure of 38 per cent (quoted in Abdullah *et al.*, 1976, p.214). The 1977 Land Occupancy Survey (LOS) reports that about 29 per cent of the households did not own any land other than homestead land (BBS, 1986b, p.69; Cain, 1983, p.158). However, this figure is likely to be less reliable, for the sample is small and may contain a high sampling error. The 1983–4 Census reports that 28.2 per cent of the households do not own any land other than homestead land (BBS, 1986b, p.69). However, the Census reports a very high percentage of farms (more than 70 per cent) in the small-farm category (less than 0.40 ha). As Cain (1983, p.154) argues, the lower end of the distribution is a sensitive indicator of change. There seems to be a process of polarization whereby the near-landless are dispossessed and join the ranks of the landless, while small farmers in turn become near-landless. Thus increasing landlessness and near-landlessness is apparent in Bangladesh, and there is considerable evidence of distress selling of land and other assets (see, for example, Alamgir, 1980, p.159 *et seq.*).

Consider now changes in the distribution of use of important components of new technology, namely, irrigation and HYVs. This may give some indication of the changing fortunes of landless and near-landless (and smaller farmers). Farm-level studies (e.g. I. Ahmed, 1981; Alauddin and Tisdell, 1988e) indicate that degree of access to irrigation is a key determinant of HYV adoption. According to Lipton (1985), in unirrigated and unreliably rainfed areas there is no positive association between amount of land owned and operated (between 0 and 2.4 ha) and poverty risk. Tiny amounts of well-watered land reduce that risk. Thus increased inequality in the distribution of irrigated land is likely to increase inequality in the distribution of incremental output.

As reported in BBS (1981, pp.41, 45; 1986b, p.82), a higher percentage of larger farms, as compared to small and medium farms, have access to irrigation. The gap in the access to irrigation has increased over time. For instance, in 1976–7, 32.7 per cent of large farms had access to irrigation compared to 31.1 and 29.4 per cent respectively of the medium and small farms. The corresponding figures for 1983–4 respectively are 57.9, 53.4 and 38.7. While differences in these percentages are not substantial in the first period, they are in the subsequent period. The intensity of irrigation (irrigated area as percentage of operated area) and that of HYV cultivation (total area under HYVs of different crops of rice and wheat as a percentage of operated area) appear to be inversely related to farm size. These intensities seem to decline with increase in operated area. Between the two census years, 1977 (BBS, 1981, pp.42–5) and 1983–4 (BBS, 1986b, p.81; 1986c, p.111) the percentage of operated area irrigated and planted with HYVs rose significantly for all classes of farmers. As of 1983–4, nearly 20 per cent of area operated by small farmers was under irrigation, compared to 17 and 16 per cent respectively for medium and large farmers, compared to 13, 10 and 9 per cent in 1977.

However, when one considers the overall distribution of area irrigated and area planted with HYVs, it can be seen that disproportionately higher percentages of these areas are under the command of the relatively larger farms. Employing agricultural census data for 1977 and 1983–4 indices of concentration of critical ingredients of the new technology are constructed. These indices are the same as employed in Chapter 7. The indices of concentration calculated for the distribution of irrigated and HYV area in 1983–4 were 0.506 and 0.439 respectively. An inter-temporal comparison suggests a trend towards greater concentration of these elements (0.355 and 0.310 for irrigated and HYV area in 1977 respectively). A word of caution, however, is warranted. As mentioned earlier, the under-reporting of small farm households in the 1977 sample census is likely to underestimate the concentration ratios and, therefore, overestimate the extent of their increase by 1983–4. Nevertheless, a high degree of inequality is clearly indicated by the 1983–4 data and an increasing trend cannot be ruled out. It is also noticeable that irrigated area is more unevenly distributed than HYV area. This is because not all HYVs are irrigated. For instance, around a third of total HYV wheat area is irrigated and the HYVs of *aus* and *aman* (*kharif*) rice are primarily rainfed.

Let us turn now to the pattern of distribution of ownership of livestock and poultry resources. Consistent with the pattern of distribution of land and other resources, the distribution of these resources is quite uneven. As of 1983–4 (BBS, 1986b, pp.83–5), while most of the medium and large sized farms reported having bovine animals (principally used as draught animals), only 55 per cent of those in the small-farm category did so. For sheep, goats and poultry, the contrast seems to be less striking. Because of differences in definition and scope in various censuses, inter-temporal comparison may not seem to be meaningful. However, a significant decline in the area under fodder crops and virtual disappearance of pasture land (BBS, 1986b, p.101) have made it more difficult for the small farms to maintain bovine populations. As BPC (1985, pp.IX–45) points out, 'fodder supply was adversely affected by food production, partly due to the decline in grazing land, but mainly due to the shift from long-stem to short varieties of rice cultivation'. The unevenness in the distribution of livestock affects smaller farms more severely than the larger ones in terms of cost of cultivation (Gill, 1981, p.14).

The preceding discussion may be summed up as follows:

1. There exists a high degree of inequality in the distribution of land and there is a trend towards greater concentration of land holdings. More interestingly, recent census reports indicate a greater degree of inequality in the distribution of operated holdings (owned + leased in − leased out) than that in owned holdings. This may imply that the access of owners of small areas of land to farming of land is becoming gradually more restricted.
2. Distribution of areas under irrigation and HYVs indicates increasing concentration.
3. On a static cross-sectional basis draught power is unevenly distributed.
4. Incidence of landlessness is increasing over time. Furthermore, there is an increasing trend in the proportion of the near-landless who border on landlessness and eventually become landless.

8.3 CONCENTRATION OF RESOURCES: FARM-LEVEL EVIDENCE

Let us now consider some farm-level evidence relating to distribution of land and other resources and the incidence of landlessness. The discussion in this section is based on information collected during a field survey on the 1985–6 crop year in two Bangladeshi villages:

Ekdala and South Rampur, reported in Chapter 7. The survey includes information on 58 land-owning households in each of the two villages, as well as on landless households. For the purpose of the survey, landless households included those who did not own any cultivable land but might or might not have any homestead land. Using a separate questionnaire, information on 18 and 24 landless households was collected respectively from South Rampur and Ekdala. The corresponding total numbers were reported to be 29 and 95 respectively.

8.3.1 Distribution of Land and Non-land Resources

An analysis of the data on land ownership and distribution indicates a significant concentration of land holdings in both villages. Smaller farmers, despite being the vast majority of the households, own and operate a relatively minor proportion of the total farm area. The values of concentration ratios clearly support this. For Ekdala, owned and operated holdings had coefficients of 0.553 and 0.500 respectively. The corresponding values of the concentration ratios for South Rampur were 0.415 and 0.387. The pattern of land distribution embodies, on both counts, a higher degree of inequality in Ekdala than in South Rampur. On the basis of ownership, the distribution in Ekdala is more unequal than that for Bangladesh as a whole. On an operational basis, the opposite seems to be the case.

Consider the observed pattern of distribution of HYVs and related innovations in these villages. Concentration ratios for area under all HYVs taken together and irrigated area in the two villages for the survey year (1985–6) show a differing pattern of distribution. While the Ekadala data produced concentration ratios of 0.441 for all HYVs and 0.423 for irrigated area, the corresponding South Rampur values were 0.319 and 0.323 respectively. The distribution of these innovations is less unequal than for land resources in the two villages, and the one for Bangladesh's overall distribution of area under HYVs and irrigation reported earlier in the chapter.

To what extent do farm-level data support the pattern of increasing concentration of irrigated and HYV areas of Bangladesh as a whole? In an earlier study (Alauddin and Tisdell, 1988f) we have addressed this issue in greater detail, but some results are worth reporting in this chapter. For that purpose we have calculated an index of concentration (mentioned earlier) for areas under HYVs and irrigation in the two villages for the 1980–1 to 1985–6 period. The results show a contrasting pattern of distribution of certain elements of the new agricultural technology for the two villages. For Ekdala there is an increasing tendency towards concentration for all innovations. Between 1980–1 and 1985–6 concentration ratios for HYVs and irrigated area increased respectively by 58 per cent (from 0.278 to 0.441) and 19 per cent (from 0.356 to 0.423). For South Rampur on the other hand, the concentration indices remain much the same with a weakly declining tendency.

The differential behavioural pattern of concentration ratios may be the result of two factors: (1) differing intensities of HYV adoption – the intensity of adoption in Ekdala is far below that of South Rampur, where a more stable pattern seems to have been in existence for some time; (2) differential pattern of land ownership distribution – South Rampur has a more equitable (Gini ratio = 0.415) land ownership distribution pattern than Ekdala (Gini ratio = 0.553). Indeed it is possible that, after the adoption of HYV innovation, concentration ratios tend at first to increase and eventually come to relatively stationary (equilibrium) values. So at least initially there appears to be some increase in dualism (see Yotopoulos and Nugent, 1976, p.238). Transmission of further income inequality through a greater concentration of innovations in areas characterized by a greater degree of inequality in the distribution of land may not be ruled out (Hayami and Ruttan, 1984, p.49).

Similar inequality can be observed in the distribution of bovine animals, sheep, goats and poultry. For instance, a higher proportion of farmers operating larger farms own these resources. While for chickens, and sheep and goats, the differences do not appear to be all that significant, for bovine (draft) animals the contrast is striking. In Ekdala, whereas 87 per cent of large farms and 91 per cent of the medium-farm households reported having bovine animals, only 47 per cent of the small-farm households reported so. One can also note inter-village differences in the percentage of farmers possessing these animals. In general (sheep and goats excepted), a higher percentage of South Rampur farmers owned these resources than in Ekdala. In terms of concentration ratios, the Ekdala sample displays a greater degree of inequality in the distribution of bovine animals compared with South Rampur (cf. concentration ratios 0.553 and 0.493). The South Rampur sample exhibits a higher degree of inequality in the distribution of bovine animals than for the distribution of land.

8.3.2 Incidence of Landlessness: Some Characteristics

The two villages differ significantly in incidence of landlessness. Ekdala, at the time of conducting the survey, had an estimated 90 landless households out of a total of 256, while the corresponding South Rampur figure was 29 out of a total of 165. Thus the incidence of landlessness in Ekdala is twice as high as that in South Rampur (over 35 per cent in Ekdala against nearly 18 per cent in South Rampur).

Table 8.1 sets out relevant information about landless households of Ekdala and South Rampur. In both villages there is a very high percentage of landless households who did not inherit any land. Some who did became landless because of poverty and medical treatment. The vast majority of the landless householders inherited landlessness.Two-thirds of the landless households of Ekdala work as agricultural labourers. The corresponding figure for South Rampur is much lower, 39 per cent. A higher percentage (83.33) of the South Rampur landless, compared to those in Ekdala (62.50), did not have any subsidiary occupation. Thus the occupational pattern in Ekdala is slightly more diversified than in South Rampur. A smaller percentage (33.33) of the Ekdala landless sharecrop land than in South Rampur (50.00). However, the amount of sharecropped land per household in Ekdala is about three times as high as that in South Rampur (0.50 and 0.17 ha respectively). The Ekdala landless seem to be better endowed with bovine and other animals and poultry. A higher percentage of households in Ekdala reported having these resources. A third of the Ekdala households owned no homestead land, while all households in South Rampur had such land. However, the average area of homestead land in Ekdala is more than twice that in South Rampur.

The pattern of distribution of land and incidence of landlessness in Ekdala seem to be more consistent with Bangladesh as a whole than does the pattern in South Rampur. The degree of concentration in other resources in Ekdala also seems more in line with the aggregate scenario than does that of South Rampur. We do not have any data to indicate whether landlessness is increasing over time in the two survey villages. However, the process by which small farmers in Ekdala and South Rampur became landless (e.g. distress selling, medical treatment, etc.) seems to suggest, as demographic pressure and institutional factors (e.g. law of inheritance) reduce the size of

Table 8.1   Characteristics of landless households in two Bangladeshi villages: Ekdala and South Rampur, 1985–6

Characteristics	Ekdala[1]		South Rampur[2]	
	Number	Percentage	Number	Percentage
1. Households not inheriting any land	22	91.67	12	66.67
2. Father landless	11	45.83	11	61.11
3. Mode of landlessness				
a. Inherited	16	66.67	12	66.67
b. Sold land because of poverty, medical treatment, etc.	2	8.33	6	33.33
c. Other reasons (including migration from other regions of Bangladesh and India)	6	25.00	–	–
4. Major occupation				
a. Agricultural labour (including day labour)	16	66.67	7	38.89
b. Others (e.g. rickshaw-pulling, petty trading, service)	8	33.33	11	61.11
5. Subsidiary occupation				
a. None	15	62.50	15	83.33
b. Agricultural labour (including day labour)	4	16.17	1	5.56
c. Others (e.g. riskshaw-pulling, petty trading, service)	5	20.83	2	11.11
6. Incidence of sharecropping				
a. Households operating land under sharecropping	8	33.33	9	50.00
b. Sharecropped land per landless household (ha): Ekdala: 0.50 South Rampur: 0.17				
7. Households reporting:				
a. Bovine animals ·	5	20.83	1	5.56

Table 8.1 continued

Characteristics	Ekdala[1]		South Rampur[2]	
	Number	Percentage	Number	Percentage
b. Goats/sheep	5	20.83	1	5.56
c. Poultry birds	16	66.67	10	55.56
8. Homestead land				
a. Households owning homestead land	16	66.67	18	100.00
b. Homestead land (ha)				
Ekdala: 0.05				
South Rampur: 0.02				

[1] Total number of landless households interviewed in Ekdala = 24.
[2] Total number of landless households interviewed in South Rampur = 18.

holding, that adverse circumstances are more likely to force small owners to sell their land. This circumstantial and statistical evidence from other areas of Bangladesh (e.g. Atiur Rahman, 1982) indicates that landlessness may be on the increase.

8.4 MARKET AND NON-MARKET INCOME OF BANGLADESHI RURAL POOR

8.4.1 *A Priori* Considerations

While the rural poor (landless and near landless) appear to earn most of their income from wages, they supplement this income by renting or purchasing natural resources or by having free access to these. But with the introduction of new technology, extension of the market and population increases, the scope for supplementing wage income has dwindled at the same time as landlessness or near-landlessness has risen (cf. Repetto and Holmes, 1983) and this has added to rural poverty and income inequality as well as poverty risk. Various mechanisms are at work: the price of scarce resources such as irrigation water tends to rise, and access to these tends to become more limited, to the detriment of the landless and near-landless, common access or inexpensive access to resources may be lost, and unfavourable spillovers affecting common property resources may intensify. Let us consider each of these matters.

With the introduction of new technology to rural areas, the rural poor may be disadvantaged because of increased demand for natural resources to satisfy the supply of inputs complementing this technology. When the price of natural resources is forced up, only those capable of adopting the new technology find it worthwhile to buy (hire) natural resources on any scale. The adoption of new technology may be limited or relatively limited to the wealthier members of the farming community because of scale, risk, finance availability and other factors. Consequently the price of a resource may rise, an individual can no longer afford to purchase it and his income falls. But the richer individual can purchase the resource and his income may increase in absolute terms. In any case the differential adoption of technology (the pattern which was outlined in the previous chapter) may increase income inequality. In addition, the poor may be forced by the changed circumstances to try to obtain full-time labouring work. As Bardhan (1984, p.189) argues, 'increased dependence on purchased inputs and privately controlled irrigation is driving some small farmers with limited access to resources and credit out of cultivation and into crowding the agricultural labor market'.

The possible impact of the pricing or higher pricing of an essential resource such as irrigation water can be illustrated by Figure 8.1. Suppose that prior to the introduction of new technology the price per unit of the essential resource is P_1. Given that the marginal physical productivity of its use on a small farm or land holding is indicated by the curve identified by MPP_A, the small farmer (farmer A) maximizes his surplus by purchasing x_1^A of the resource per unit of time. Similarly, the large farmer (farmer B) purchases x_1^B of the resource, given that the curve identified by MPP_B is the marginal physical productivity of its use. The small hatched triangle on the left-hand side represents the net surplus obtained by the near-landless farmer, whereas that on the right-hand side is the surplus of the large farmer. Now suppose that, with the introduction of the new technology, the demand for using the resource rises so its price increases to P_2 per unit. By adopting the new technology, farmer B raises his marginal physical productivity curve to MPP_B' and obtains a surplus shown by the dotted area. By contrast, if A is unable to adopt the new technology, he finds it uneconomic to to purchase the essential resource at the price P_2 and is forced out of farming. Even if MPP_A happened to intersect the price line P_2 it is clear that the

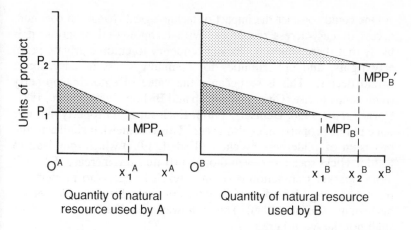

Figure 8.1   Income inequality and landlessness caused by rising natural resource scarcity

income of the small farmer would be reduced compared to the pre-new-technology situation, and income is likely to be more un-evenly distributed.

The extent to which the income distribution is likely to become more unequal may be even greater than indicated by changes in the surpluses. Larger farmers may be the sole suppliers of essential resources such as irrigation. Consequently they appropriate the gains from its sale or direct use. For example, the income of farmer B prior to new technology may be greater than indicated above by $P_1x_1^A$ (his charges to farmer A for resource X) plus $P_1x_1^B$ (market value of own use of the resource). After the introduction of the new technology the income of the large farmer increases to an amount equal to the dotted area plus $P_2x_2^B$, whereas the farm income of A becomes zero.

A similar diagram (Alauddin and Tisdell, 1989d) can be drawn to show how the introduction of new technology may cause free access by the landless or near-landless to a resource to be discontinued and the denial of access to them by the larger farmers at any price which the small farmers would find economic. Note, however, that purely competitive markets (because of locational and transport factors) do not typically exist in Bangladesh in such resources (large farmers can, for instance, be water monopolists) and actual costs may be involved in extracting the natural resource, e.g. pumping irrigation water. Nevertheless, these considerations do not change the substance of the argument.

One could consider the impact of technological change on common-access or easy-access resources. With technological progress, it is likely that there would be greater tendency to enforce private property rights, and opportunities for common access or easy access would decline. This is so because the value of enforcing property rights tends to rise (Ciriacy-Wantrup and Bishop, 1975; Runge, 1981; Cornes *et al.*, 1986; Demsetz, 1967). Enclosures of property that was once public property may also occur. This is somewhat similar to the provision of wilderness by clubs (Tisdell, 1984) which may lead to Kaldor-Hicks social loss even though private gains increase. Furthermore, market transaction costs may result in access to property or resources being denied once they become valuable to the owner for his own use (Posner, 1980). This is, however, a different issue and we shall not discuss it here.

In addition, unfavourable externalities on common property may occur. Owing to technological progress it may become more profitable to clear land, increase its drainage, avoid leaving it fallow for long and cultivate it more intensively. This may result in loss of wildlife and wild fruits that may be gathered by the poor. Also, there may be adverse spillovers on fish population, etc., from greater use of pesticides and chemical fertilizers. These effects may have their greatest impact on the poor who tend to be gatherers rather than farmers. Also, loss of access to 'free' fuel (firewood) and to thatching and housing materials can be expected.

Clearly the whole matter is very complicated. If the poor lose access to (free or inexpensive) natural resources, their economic position depends upon whether there is a corresponding increase in the demand for their labour in the market. As the poor lose easy access to natural resources, they are likely to become more dependent on hired employment.

8.4.2 Some Empirical Evidence

If one recalls the evidence in Chapter 6 of increase in functional (hired or family) demand for labour, it becomes clear that family labour has made more significant gains than hired labour, even though the demand for hired labour has also increased. Assuming that increased demand for hired labour has to some extent compensated the landless for their restricted access to (free or inexpensive) resources, an important question arises: To what extent have real agricultural wages increased in recent years? The welfare of the rural poor is critically dependent on wage entitlements. Table 8.2 sets out data on nominal wage rate, cost of living index and real wages for Bangladeshi agricultural workers during the 1969–70 to 1985–6 period. Nominal wages (NWAGE) have increased by nearly a factor of 10. We have used three different indices, country cost of living for industrial workers (INDCPI), rice price index (RICEPIND) and price index of the food component of agricultural products (FOODCPI). All the three indices of cost of living show considerable increasing trends. This implies that there is little underlying trend in real wages. Real wages (RWAGE1, RWAGE2, RWAGE3, RWINDEX1, RWINDEX2, AND RWINDEX3) dropped sharply during the early 1970s and remained depressed till the early 1980s. Some increasing trend is apparent in the last two to three years. This is also illustrated with the aid Figure 8.2.

In this respect, our findings are consistent with those of A.R. Khan (1984) but one needs to be reminded of the limitations of these indices. They are based on averages, and conceal regional and seasonal variations in wages. Nevertheless, they show sudden drops in real wages to drastically low levels in adverse natural and weather conditions (e.g. flood of 1974, drought of 1979). These sharp declines in real wages imply a significant decline in food-entitlements and a greater incidence of poverty (see A.K. Sen, 1981).

Although this may not strictly apply to large parts of Bangladesh, natural environments form resources to fall back on in bad times. They are like a security device. Lipton (1985) discusses these issues. Clarke (1971) specifically observes this in relation to New Guinea. In the absence of these cushions the poor are likely to suffer greatly during times of economic stress or economic difficulty.

In Ekdala and South Rampur, households dependent on on-farm employment reported serious economic risk problems. In abnormal years when there is a serious drought or a flood, not only do real wages fall drastically, but also employment becomes more limited. Opportunities for earning supplementary incomes by raising cattle and other animals and poultry become increasingly restricted. Furthermore, the growing of vegetables for domestic consumption or for sale at critical times is becoming more difficult because of scarcity of water. Landless labourers and small-farm households in the Ekdala area reported having difficulties in getting water from privately owned tanks or ponds for watering vegetables or seasonal fruit plants. It is not that the water from these sources has been priced, but access to it has been restricted for two reasons: (1) water as a

Table 8.2 Daily wages of agricultural labourers: Bangladesh, 1969–70 to 1985–6

Year	NWAGE	INDCPI	FOODCPI	RICEIND	RWAGE1	RWAGE2	RWAGE3	RWINDEX1	RWINDEX2	RWINDEX3
1969–70	2.96	100.00	100.00	100.00	2.96	2.96	2.96	100.00	100.00	100.00
1970–1	3.13	104.00	105.00	98.47	3.01	3.18	2.98	101.68	107.39	100.71
1971–2	3.38	108.00	122.00	130.01	3.13	2.60	2.77	105.73	87.83	93.60
1972–3	4.72	193.00	187.00	198.48	2.45	2.38	2.52	82.62	80.34	85.27
1973–4	6.69	268.00	257.00	263.87	2.50	2.53	2.60	84.33	85.65	87.94
1974–5	9.05	431.00	508.00	531.58	2.10	1.70	1.78	70.94	57.52	60.19
1975–6	8.82	365.00	343.00	324.64	2.42	2.72	2.57	81.64	91.79	86.73
1976–7	8.93	354.00	323.00	288.48	2.52	3.09	2.76	85.22	104.58	93.40
1977–8	9.44	419.00	377.00	369.26	2.25	2.56	2.50	76.11	86.37	84.59
1978–9	10.88	458.00	430.00	396.18	2.38	2.75	2.53	80.25	92.78	85.48
1979–80	12.46	526.00	528.00	533.89	2.37	2.33	2.36	80.03	78.84	79.72
1980–1	13.97	568.00	545.00	460.80	2.46	3.03	2.56	83.09	102.42	86.60
1981–2	15.48	656.00	624.00	577.74	2.36	2.68	2.48	79.72	90.52	83.81
1982–3	17.05	684.00	633.00	643.90	2.49	2.65	2.69	84.21	89.46	91.00
1983–4	19.58	761.00	725.00	698.28	2.57	2.84	2.70	86.92	95.97	91.24
1984–5	24.54	356.00	817.00	783.91	2.87	3.13	3.00	96.85	105.76	101.47
1985–6	29.53	941.00	911.00	753.13	3.14	3.92	3.24	106.02	132.46	109.51

Note: NWAGE is nominal wage (taka/day); INDCPI, FOODCPI, RICEIND respectively refer to indices of cost of living of industrial workers, food and rice prices. RWAGE1, RWAGE2 and RWAGE3 are real wages (taka/day) and are derived by deflating NWAGE by INDCPI, RICEIND and FOODCPI respectively. RWINDEX1, RWINDEX2 and RWINDEX3 are indices RWAGE1, RWAGE2 and RWAGE3. All indices are based on 1969–70 = 100.

Source: BBS (1979, p.386; 1984b, p.660; 1986d, pp.17, 21).

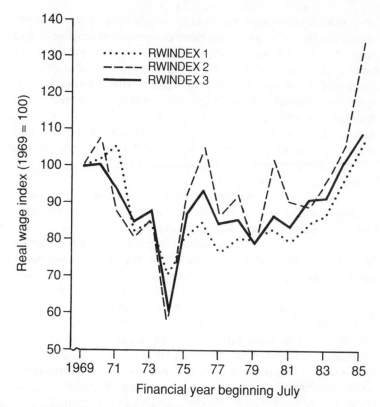

Figure 8.2 Trends in daily real wages of agricultural workers using various price deflators: Bangladesh, 1969–70 to 1985–6

resource is valued more highly by the owners; (2) there are transaction costs involved in monitoring its sales if it were priced.

Because of gradual destruction of permanent fruit trees to clear land for food production, the gathering of fruits in hard times is even more difficult. Also, loss of shrubs and trees results in loss of leaves and twigs used as cooking fuel as well as firewood. Thus the opportunities to fall back on reserves of natural resources, especially during difficult times, seem to be disappearing with increased population and greater penetration of market forces following new technology.

Most farmers in South Rampur reported serious shortages of available grazing land for cattle, and many reported a decrease in livestock population over the years. A chi-square test indicated that the grazing problem was independent of farm size (chi-square value significant at a probability level = 0.1806). Even in cases where there were not apparent declines in cattle population, some farmers reported having to restrict their number because of grazing or fodder problems. For Ekdala, however, a chi-square test indicated significant variation of grazing problems across farm size (chi-square value significant at a probability level of 0.0022). This is probably because larger-farm households could provide adequate cattle feed from rice straw and other crop or crop by-products, e.g. *khesari*, sugar cane leaves, etc. The South Rampur farm households did not have a greater degree of freedom in that respect and because of higher intensity of cultivation of (short-stemmed) HYVs, cattle feed problems were even more serious.

It is interesting to consider whether the farmers believe that their economic situation is more risky or uncertain compared with the past. For Rampur most farmers thought that this was so. A chi-square test indicated that the null hypothesis of independence across farm size could be accepted with a high degree of confidence (chi-square value significant at a probability level of 0.2670) against an alternative of variation of poverty risk across farms. For Ekdala, however, the alternative hypothesis could not be accepted so confidently (chi-square value significant at a probability level of 0.0593). This suggests that small-farm households in particular feel they are at a greater risk than before. The inter-village difference may probably be because of difference in density of population, variations in size of small and large farms and perhaps diversity of cropping patterns. However, in both villages farmers appear to feel that they are at greater economic risk than ever before.

8.5 CONCLUDING OBSERVATIONS

An analysis of the distribution of land and the trends in landlessness and near-landlessness indicates increasing dependence on wage employment for subsistence by the rural poor. However, seasonality of agricultural employment and prices of wage goods (primarily food) are a critical determinant of the seasonal dimension of poverty (Chaudhury, 1981a; Clay, 1981). Furthermore agricultural wages, being close to subsistence level, provide little scope for carry-over into periods of slack agricultural activity (A.R. Khan, 1984). Even though there is some evidence of a slightly increasing trend in real wages, much of its effect on rural poverty is neutralized because of seasonality in employment and real wages.

The income of the rural poor in Bangladesh (like that in most LDCs) consists of two components: exchange and non-exchange. Even though in normal times the exchange component is important to the rural poor, during slack periods of employment the non-exchange sources of income become more important. This assumes even greater significance in *abnormal* years when both real wage and employment levels fall to dismally low levels. With rapid population growth, depletion of natural resources and greater penetration of technological and market forces, the cushioning effect of access to natural resources, in adverse circumstances, on the rural poor has become more limited and their income security has been undermined (Mosharaff Hossain, 1987).

It seems clear from an analysis of both agricultural census and survey data that the opportunities to fall back on reserves of natural resources in times of stress or economic difficulty are gradually disappearing. Recent studies of rural poverty and income distribution in Bangladesh have not taken these factors into account. It is argued in this chapter that without taking these factors into account a realistic picture cannot be obtained of the process underlying poverty and inequality.

Finally, the limited empirical basis makes generalizations risky and, therefore, the conclusions reached must be considered as tentative. Nevertheless, the gravity of the problem of resource depletion and restricted access to resources by the rural poor in a country like Bangladesh, which is plagued by natural disasters of one kind or another, can hardly be overemphasized. While one should not despair, there is cause for concern and further research is warranted.

9 The 'Green Revolution' and Variability of Bangladeshi Foodgrain Production and Yield: An Analysis of Time Series Data

9.1 INTRODUCTION

In the survey discussed in the previous chapter, Bangladeshi farmers claimed that their income had become more uncertain and risky since the Green Revolution. While there may be a number of reasons for this, one possible factor could be that the yield and production of foodgrains have become more variable. This possibility will be investigated in this chapter using time series data.

Foodgrains are the most important wage goods in the LDCs, including Bangladesh, and the income elasticity of demand for food is very high, probably of the order of 0.60 or higher (Johnston and Mellor, 1961, p.572; Mahmud, 1979, p.65). In the short run, changes in relative food prices materially alter the real incomes of individuals on low monetary incomes (Mellor, 1978, p.1; 1985). With technological change in the production of foodgrains (the Green Revolution), the supply of foodgrains may have become more variable and the prices of foodgrains more unstable. For example, it is commonly believed that yields from HYVs are more variable than those from the local varieties (see, for example, Mehra, 1981). In the LDCs this can have important welfare consequences for low-income earners and increase fluctuations in incomes received by grain producers. Consumers and producers can suffer welfare losses through price instability resulting from yield and production instability. This may also lead to an increase in risk for farmers and may hinder farm-level innovation adoption, and hence arrest the pace of national foodgrain output. Furthermore, increased yield variability in foodgrains or export crops can lead to destabilization in such macro variables as national income, employment and balance of payments (Hazell, 1986, pp.41–3).

As discussed in Chapter 2, foodgrain production is of critical importance to the agricultural economy of Bangladesh as reflected in percentage of cultivated area allocated to rice and wheat (BBS, 1986d). For centuries, Bangladesh was self-sufficient in food. In recent decades, however, it has become a net importer of food and its food imports have been growing. Imports as a percentage of total available foodgrains for consumption have increased from less than 2 per cent in the early 1950s to well over 10 per cent in recent years (Alauddin and Tisdell, 1988b). At the same time, agricultural production in Bangladesh remains erratic because of considerable year-to-year climatic variability (BBS, 1985b, p.303). Fluctuations in domestic grain production are usually compensated for by fluctuations in grain imports (Alauddin and Tisdell, 1987) and this periodically strains the already scarce foreign exchange reserves of Bangladesh and makes development planning more difficult.

As mentioned in previous chapters, the last two decades have witnessed the introduction of Green Revolution technology. This has resulted in a steady improvement in overall agricultural production, due mainly to a moderate increase in the production of rice and a spectacular boost in the production of wheat (Wennergren *et al.*, 1984, p.75; Alauddin and Tisdell, 1987).

In the light of recent developments in the input as well as in the output sides of Bangladeshi foodgrain production, this chapter analyses data on the variability of Bangladeshi foodgrain production and considers the possible role of the new technology (the Green Revolution) in moderating or accentuating fluctuations in production and yield. In this discussion, the main focus is on the aggregate supply of foodgrains (rice and wheat). This chapter first of all briefly reviews the relevant literature and contends that the methodologies adopted by the authors of some recent studies (e.g. Hazell, 1982; Mehra, 1981) suffer from serious shortcomings, can lead to misleading conclusions and be a source of faulty policy advice. It then presents and applies an alternative methodology in order to determine and outline trends in the variability of Bangladeshi foodgrain production and yield over time, first determining sub-periods on observed mathematical grounds, and subsequently on technological considerations. Factors are then considered which may explain the observed underlying trends in the variability of foodgrain production in Bangladesh.

9.2   A BRIEF REVIEW OF THE LITERATURE

The question of the stability and adaptability of crops has been discussed at theoretical and empirical levels. Studies include those by Evenson *et al.* (1979), Finlay and Wilkinson (1963) and Tisdell (1983b). Evenson *et al.* (1979) point to the need to draw a distinction between (a) *stability* of a genotype, that is, its changing performance with respect to environmental factors over time, and (b) *adaptability*, that is, its performance with respect to environmental factors that change across locations. Evenson *et al.* express concern that new High Yielding Varieties (HYVs) of crops could increase yield variability in developing countries and recommend more research into crops with a view to reducing such variability.

Recent in-depth studies of Indian agriculture (Mehra, 1981; Hazell, 1982; 1984; 1985) found evidence of increased instability in agricultural production following the introduction of modern agricultural technology.[1] Parthasarathy (1984, p.A74) indicates a higher degree of variability following the Green Revolution in the Indian state of Andhra Pradesh. He further claims that greater yield instability is positively associated with districts experiencing higher agricultural growth rates. Hazell (1982, p.10) goes so far as to conclude that 'production instability is an inevitable consequence of rapid agricultural growth and there is little that can be effectively done about it'. One needs to be reminded, of course, that despite apparent similarity in conclusions Mehra (1981) and Hazell (1982) differ in an important respect. Mehra hypothesizes a causal link between the new technology and increased production and yield instability. Thus while Mehra attributes most of the production variation to yield instability, Hazell (1982; 1985) attributes production variability to yield variability as well as a reduction in the offsetting patterns of variations in yields between crops and regions.

While the studies by Hazell and Mehra are substantial, in the view of the present authors they are subject to two methodological limitations. First, these studies measure production stability or lack of it around a line of 'best fit'. As Ray (1983, pp.462–3) points out, 'any inference regarding changes in the pattern of growth and instability in production will be greatly influenced by the choice of mathematical function, the selection of which cannot be left alone to the statistical criteria of best fit'.[2] Furthermore, while their studies compare the variability of one period with that of another, they do not consider whether variability itself shows any tendency to increase or decrease within a period of specified duration. Ray's own study suffers from similar inadequacies in that it compares the 'instability of production over different periods under the assumption of a stationary mean and variance in the year to year changes in the components of production index' (p.463).

Secondly, both Hazell and Mehra assume arbitrary cut-off points. Furthermore, they do not seem to have a consistent rule for dropping observations for 'unusual' years. Probably both Mehra and Hazell were justified in dropping observations relating to 1965–6 and 1966–7 because of severe drought during those years. As Hazell (1982, p.13) points out, 'catastrophes of this kind are sufficiently rare and severe that they can be considered as separate phenomena from year to year fluctuations'. Mehra (1981, p.10) argues that, 'the mid-1960s witnessed two drought years 1965–6 and 1966–7 of such unusual severity as to significantly alter the variance of any period in which they are included, thus casting doubts about the validity of their conclusions'.

On closer examination of the foodgrain production data presented by Sawant (1983, p.476) one can identify two worst years during the period 1967–8 to 1977–8 which corresponds to the second period designated by Mehra and Hazell. In 1972–3 Indian foodgrain production dropped by over 8 million metric tons (8 per cent) from the previous year's production. It was even worse in 1976–7 when the decline was 10 million metric tons (over 8 per cent) from the production of 1975–6. Apart from 1965–6 and 1966–7, no other year between 1950–1 and 1977–8 saw such an absolute decline in foodgrain production in India. The seriousness of the problem can also be seen if one considers the yields of the two major foodgrains (rice and wheat) which together constitute over 70 per cent of all foodgrains during the period 1967–8 to 1977–8 (Hazell, 1982, p.13). Using data from Joshi and Kaneda (1982, p.A3), one can identify a few bad years in terms of yields per hectare. During 1972–3 wheat yield fell by 109 kg per ha (over 8 per cent) from the 1971–2 level of 1380 kg. In 1973–4 it dropped by another 8 per cent. For rice, 1972–3 was not as bad as it was for wheat. But during 1974–5 rice yield dropped to 1045 kg per ha (over 9 per cent) from the previous year's 1151 kg per ha. The worst year for rice during the second period must be 1976–7 when rice yield fell to 1088 kg from 1235 kg per ha in 1975–6 (more than 9 per cent). Overall, of course, 1972–3 and 1976–7 remain the two worst years during the second period. To be consistent, one would have expected these two years to be dropped from the analysis of the second period. In that case, one would perhaps end up with a different picture from those emerging from the studies by Mehra (1981) and Hazell (1982).

It seems likely that the findings of both Mehra and Hazell are sensitive to changes in cut-off points and to their decisions to delete certain observations. This gains some support from a recent study by Hazell (1985). In that study, Hazell compares instability in world cereal production between two periods, namely, 1960–1 to 1970–1 and 1971–2 to 1982–3, and also examines instability in cereal production in different regions of the world, e.g. in South Asia, in India. When comparing the instability of cereal production between the two periods for India, he does not drop observations for 1965–6 and 1966–7. Nor does he drop the observations for 1972–3, 1976–7 or 1979–80 when total foodgrain production fell by a huge 22 million tonnes, i.e. about 17 per cent (Sawant, 1983, p.476). When no observations are dropped from either period, one finds that the coefficient of variation of cereal production in India decreases by 29 per cent (Hazell, 1985, p.150) during the second period (1971–2 to 1982–3) as compared to the first (1960–1 to 1970–1), whereas the earlier studies by Mehra and Hazell indicated a rise in the coefficient of variation. Thus the assumption of arbitrary cut-off points and inconsistency in deletion of observations can lead to conflicting results.

At this stage it is pertinent to mention that while Hazell (1982) and Mehra (1981) found increased variability in foodgrain production and yield following the introduction of the Green Revolution technologies, some studies also provide evidence to the contrary. In an earlier study, Sarma and Roy (1979) found that the coefficient of variation of India foodgrain production declined from 14 per cent in the pre-Green Revolution (1949–50 to 1964–5) to 8 per cent in the period following it (1967–8 to 1976–7).[3] A recent study (Jain et al., 1986) extends the Hazell (1982) analysis for India to 1983–4 and, without dropping any observations from either period, finds that the period of the new technology (1967–8 to 1983–4) is associated with a lower production and yield variability compared to the earlier period (1949–50 to 1966–7). One further point that emerges from recent studies is that while Hazell (1986, p.16) shows that the probability of a 5 per cent fall below the trend in world cereal production may have doubled in recent years, there are wide variations between regions and commodities. For instance, the coefficients of variation of both rice and wheat production have declined in recent years (Hazell, 1986, p.18; Evans, 1986, p.2). Hazell (p.18) also claims that 'the least risky countries are those that predominantly grow rice, presumably because much of the crop is irrigated. These countries include Indonesia, Thailand, Bangladesh and Japan'. Thus more recent evidence seems to cast some doubt on the validity of the earlier Hazell contention of instability being an inescapable consequence of increased agricultural growth.

As will be seen, the analytical framework employed in this chapter does not measure instability around an arbitrarily fitted trend line, nor does it delete observations nor assume arbitrary cut-off points. However, it enables one to identify particular phases during which variability tends to increase, decrease or remain more or less stationary (Alauddin and Tisdell, 1988c). In the initial analysis, the cut-off points are decided on the basis of the patterns that emerge from the observed behaviour of variability over the years. In subsequent analysis, to examine the behaviour of variability within the periods of traditional and modern technologies and make a comparison between the two periods, cut-off points are adjusted so as to coincide with customary or accepted points of technological change. Thus both considerations are covered by the present analysis.

9.3 METHODOLOGY

Examination of the time-trends of variability within a region may be considered, employing two measures of variability: absolute and relative. Absolute variability is measured in terms of standard deviation, while the coefficient of variation provides a measure of relative variability. First of all, annual indices of relevant variables, e.g. production, were derived using the average of the triennium ending 1977–8 as the base. For yield, the weighted average of yields for the years 1975–6 to 1977–8 has been used as the base. The degree of variability can be measured in terms of deviations from a moving average of a specified period (say five years) of the relevant variables. The variance of a variable for any year is estimated as its observed variance for the five-year period up to and including the year under consideration. So the variance itself is a moving value. This enables one to get a series of values of absolute and relative measures of variation and identify particular phases during which variability tends to increase, decrease or remains more or less stationary. Using this approach and applying it to Bangladeshi time series data during the period 1947–8 to 1984–5, the changing behaviour of production and yield variability over time is briefly examined. Foodgrain production and yield indices and cropping intensity figures for Bangladesh for the years 1947–8 to 1984–5 are presented in Table 9.1.

9.4   EMPIRICAL OBSERVATIONS: BEHAVIOUR OF VARIABILITY OVER TIME

Table 9.1 sets out the absolute and relative variability of production and yield of all foodgrains. A few observations seem pertinent. First, there seems to be little time trend in both absolute and relative measures of variability. Secondly, absolute production variability seems to increase initially and then appears to stabilize, albeit with occasional minor fluctuations. Relative production variability increases at first and then seems to decline over time. Thirdly, both absolute and relative measures of yield variability show initial increases and then tend to stabilize if not decline.

These aspects come into sharper focus when one plots these observations against time. Figure 9.1 plots standard deviation (absolute variability) of production. One can visually identify two distinct phases (the first phase 1949–60, that is 1949–50 to 1960–1 and the second phase 1961–82, that is 1961–2 to 1982–3). Up to the early 1960s standard deviation of production shows a strong tendency to increase. In order to make quantitative comparison of change in behaviour of absolute production variability, three regression lines with time (T) as the explanatory variable were estimated. These are presented as Equations (1), (2) and (3) for the corresponding periods in Table 9.2. Figure 9.1 illustrates Equation (3) for the whole period. Equation (1) clearly indicates a strong time trend in standard deviation in terms of both explanatory power and statistical significance. Equation (2) shows no trend in absolute variability during the later phase. The estimates are poor both in terms of R^2 and t-value. Equation (3) provides a similar picture, even though R^2 is marginally higher and the coefficient is statistically significant. Overall there seems to be some tendency for the standard deviation to increase. This can also be seen by comparing the average values of standard deviation of production over the two periods. These are 5.49 and 6.67 respectively. Thus the absolute production variability seems to have risen to a higher level. This is probably influenced by some natural and political factors such as drought, flood and the War of Liberation in the early 1970s, which might have increased variability. But for all

Table 9.1 Mean values, standard deviations and coefficients of variation of production and yield of all foodgrains: Bangladesh, 1947–8 to 1984–5

Year	Production				Yield				Cropping intensity	
	Index	Mean	SD	CV	Index	Mean	SD	CV	Value	Mean
1947	52.75				70.74				130.19	
1948	60.06				78.79				131.01	
1949	57.78	56.64	2.79	4.93	75.44	73.50	3.80	5.17	127.69	129.21
1950	57.49	57.58	1.75	3.04	73.29	73.45	3.84	5.23	127.79	127.47
1951	55.11	58.48	3.56	6.10	69.23	72.65	2.70	3.72	129.38	129.87
1952	57.46	58.82	3.56	6.05	70.50	71.78	2.24	3.12	131.44	130.61
1953	64.56	57.32	5.37	9.36	74.80	70.20	3.39	4.83	133.08	130.44
1954	59.46	59.12	5.92	10.01	71.06	76.62	5.91	8.14	131.36	129.91
1955	50.02	59.53	5.84	9.81	65.40	73.51	5.85	7.96	126.94	129.16
1956	64.09	57.46	5.43	9.45	81.34	72.62	5.94	8.18	126.73	127.94
1957	59.51	58.86	6.80	11.55	74.96	74.40	6.65	8.94	127.68	127.40
1958	54.24	63.78	7.64	11.98	70.34	78.67	6.27	7.97	127.00	128.47
1959	66.45	65.80	8.97	13.63	79.98	80.41	8.12	10.10	128.69	129.29
1960	74.59	67.60	8.26	12.23	86.75	81.62	7.53	9.23	132.27	130.26
1961	74.21	73.13	6.05	8.27	90.02	86.29	5.84	6.77	130.81	131.59
1962	68.51	76.03	5.50	7.23	81.02	88.38	4.79	5.42	132.52	132.89
1963	81.91	77.31	5.81	7.51	93.68	88.86	4.71	5.30	133.65	133.79
1964	80.96	77.28	5.83	7.54	90.42	87.64	5.10	5.82	135.23	135.14
1965	80.97	81.29	5.14	6.32	89.14	89.86	3.71	4.13	136.76	137.55
1966	74.04	82.81	6.36	7.68	83.96	89.92	3.80	4.23	137.54	139.62
1967	88.54	85.26	7.68	9.01	92.09	90.35	3.98	4.41	144.57	142.75
1968	89.54	86.40	7.30	8.45	94.00	90.42	3.96	4.38	143.98	143.84

continued on page 188

188

Table 9.1 continued

Year	Index	Production Mean	SD	CV	Index	Yield Mean	SD	CV	Index	Cropping intensity Value	Mean
1969	93.19	87.06	5.91	6.79	92.57	90.64	3.51	3.87		150.90	143.99
1970	86.66	85.01	6.96	8.19	89.46	88.86	4.67	5.26		142.19	142.94
1971	77.38	85.57	7.52	8.78	85.08	89.21	5.19	5.82		138.28	142.20
1972	78.26	84.46	6.44	7.63	83.17	89.09	5.10	5.73		139.34	139.88
1973	92.37	87.08	9.50	10.91	95.75	90.94	6.70	7.37		140.27	139.75
1974	87.64	90.07	7.90	8.77	92.01	92.98	5.98	6.43		139.29	140.44
1975	99.76	96.00	7.95	8.28	98.70	97.55	5.30	5.43		141.58	142.72
1976	92.31	99.28	9.35	9.42	95.28	99.31	5.96	6.00		141.74	145.29
1977	107.93	102.61	6.77	6.60	106.03	101.30	4.36	4.30		150.74	148.07
1978	108.79	105.67	8.42	7.97	104.52	103.45	5.31	5.13		153.10	150.49
1979	104.24	109.64	4.15	3.78	101.97	105.55	2.71	2.57		153.18	152.92
1980	115.07	111.59	5.27	4.73	109.46	106.32	3.35	3.15		153.71	153.70
1981	112.19	113.99	6.32	5.54	105.79	108.04	4.27	3.95		153.86	153.71
1982	117.67	117.90	4.57	3.88	109.85	111.18	4.46	4.01		154.66	153.46
1983	120.81				113.12					153.16	
1984	123.77				117.66					151.90	

Notes: 1947 means 1947–8 (July 1947 to June 1948) etc. Index numbers are constructed with the average of the triennium ending 1977–8 = 100. Mean values are five-yearly moving averages of those index numbers. Standard deviations (SD) are based on the corresponding five-yearly figures. CV = [(SD)/Mean] × 100.

Sources: Based on data from sources mentioned in Table 2.1.

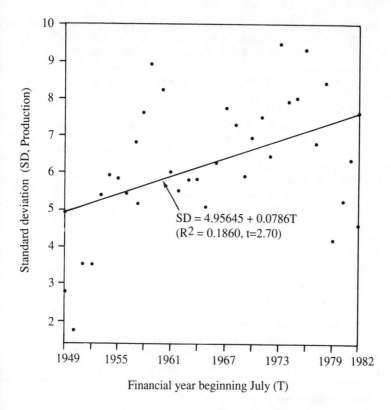

Figure 9.1   Absolute production variability: Standard deviation of aggregate foodgrain production, Bangladesh, 1949–50 to 1982–3

these factors, the average absolute production variability might be of the same order of magnitude during either phase. In any case, one would anticipate a rise in standard deviation on account of the substantial rise in the *absolute* level of production.

Figure 9.2 plots coefficients of variation (relative variability) of production against time. One can identify two similar phases in its behaviour to those for absolute variability of production. However, there is a difference, in that relative variability has declined to lower average value. It has fallen from an average of 9.01 for the first phase to 7.42 for the second. To facilitate quantitative comparisons, Equations (4), (5) and (6) were estimated by least squares linear regression and Equation (6) is illustrated in Figure 9.2. Equation (4) corresponds to the first phase and shows that variability has a strong

Table 9.2   Trends in foodgrain production and yield variability (standard deviation and coefficient of variation) for different phases divided according to Plots 1–4: Bangladesh, 1949–50 to 1981–2

Equation no.	Period	Intercept	Coefficient	R^2	t-value
Standard deviation: Production					
1	Phase 1 (1949–60)	2.22141	0.5944	0.9158	10.43[a]
2	Phase 2 (1961–82)	6.47154	0.0188	0.0073	0.38[d]
3	Entire period	4.95645	0.0786	0.1860	2.70[a]
Coefficient of variation: Production					
4	Phase 1 (1949–60)	4.31184	0.8544	0.8806	8.59[a]
5	Phase 2 (1961–82)	8.60583	–0.1126	0.1731	2.05[b]
6	Entire period	8.92703	–0.0572	0.0527	1.33[d]
Standard deviation: Yield					
7	Phase 1 (1949–60)	2.60628	0.4692	0.7720	5.82[a]
8	Phase 2 (1961–82)	4.82103	–0.0143	0.0090	0.43[d]
9	Entire period	4.92177	–0.0042	0.0009	0.17[d]
Coefficient of variation: Yield					
10	Phase 1 (1949–60)	3.85838	0.5496	0.7232	5.11[a]
11	Phase 2 (1961–82)	5.66596	–0.0657	0.1299	1.73[b]
12	Entire period	6.81343	0.0706	0.1391	2.27[b]

[a] Significant at 1 per cent level;
[b] Significant at 5 per cent level;
[c] Significant at 10 per cent level;
[d] Not significant.

tendency to increase during phase 1. The strong explanatory power and high *t*-value lend clear support to this claim. However, there is a dramatic change in the behaviour of relative variability when one considers Equation (5) which relates to the second phase and Equation (6) which relates to the entire period. Even though the signs seem to indicate a declining trend over time, one should note the poor quality of the estimates both in terms of R^2 and *t*-values. In view of this, one needs to be wary. There is no clear simple downward trend but *a fortiori* no upward trend is apparent.

Figure 9.3 depicts the behaviour of absolute yield variability over the years. Overall no time trend can be established. But two distinct phases can be identified. Absolute yield variability increases during the period up to the early 1960s (Phase 1) and falls to lower values on

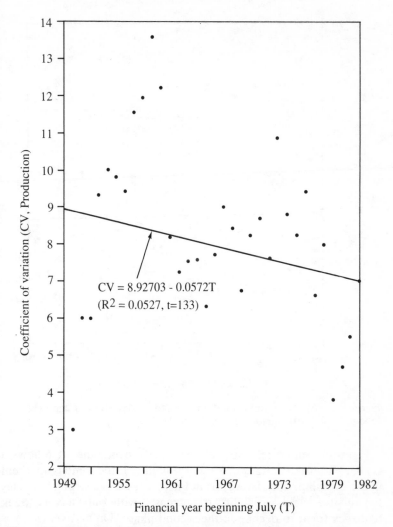

Figure 9.2   Relative production variability: Coefficient of variation of aggregate foodgrain production, Bangladesh, 1949–50 to 1982–3

average in the subsequent period (Phase 2). The average value of the standard deviation of yield falls from 5.18 in Phase 1 to 4.67 in Phase 2. The behaviour of absolute variability can be placed into a pattern by considering Figure 9.3 and Equations (7), (8) and (9) which relate respectively to the first and second phases and the entire period. Figure 9.3 illustrates Equation (9).

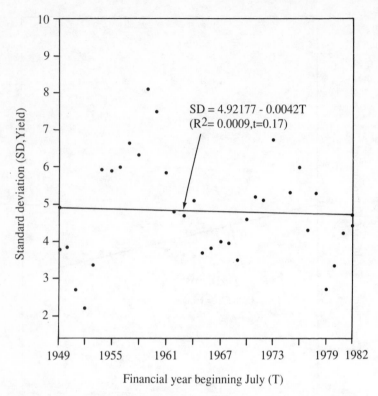

Figure 9.3   Absolute yield variability: Standard deviation of aggregate foodgrain yield, Bangladesh, 1949–50 to 1982–3

Figure 9.4 plots relative yield variability over time. It follows a similar pattern to that for absolute variability. Little trend is apparent in the overall period. However, in Phase 1 there is a strong tendency for relative yield variability to increase, with only a very weak tendency for it to decline in the second phase. On an average, it is lower during the second phase than the first phase. It falls from an average of 6.86 to an average of 4.97. Regression equations (10), (11) and (12) ((12) is graphed in Figure 9.4) indicate patterns of relative yield variability over time. These are respectively for Phase 1, Phase 2 and the entire period.

In terms of the above analysis of absolute and relative production and yield variability, there is little evidence to suggest that variability has increased in the second phase compared to that of the first phase. In fact, if one takes into account the unprecedented drought and

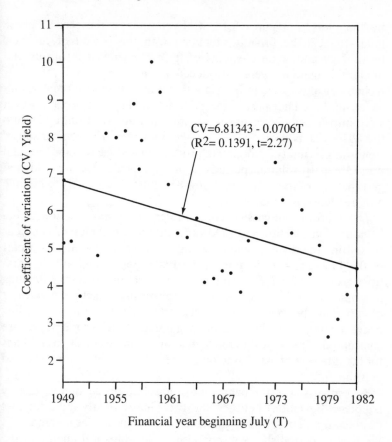

Figure 9.4   Relative yield variability: Coefficient of variation of aggregate foodgrain yield, Bangladesh, 1949–50 to 1982–3

flood of 1972 and 1974 respectively, and political factors like the War of Liberation of 1971, the second period seems to be associated with a somewhat greater production and yield stability than Phase 1. Why does the apparent break occur in variability? Mostly, it is suggested, because the second period, 1961–2 to 1982–3, corresponds to the commencement of the introduction of the new technology *to control* agricultural production. Increased irrigation and fertilizer use, followed later by the introduction of HYVs, occurred during the second phase, as distinguished here.

Nevertheless, so far the entire period has been divided into two phases only, on the basis of mathematical considerations as depicted by the plots in Figures 8.1–8.4. Even though the second phase (1961–82) can be broadly identified with the period of the new technology and some elements of it (e.g. chemical fertilizers and modern irrigation) were introduced in the early 1960s, the new technological package in Bangladesh did not assume any real significance until the later part of the 1960s when HYVs of rice and wheat were introduced. Widespread adoption did not start until well into the 1970s. Thus if one were to make inter-temporal comparison of trends in variability in foodgrain production and yield in the pre- and post-Green Revolution periods, *a priori* considerations would suggest that any of the years from 1967–8 to 1969–70 are likely to provide a more clear-cut dividing line.

The following discussion concentrates on an examination of behaviour of production and yield variability of Bangladeshi foodgrains by shifting the dividing line between the two phases to the later part of the 1960s, as may be more logical from technological considerations. As the Green Revolution is a continuous process, a dividing line based on a single year may not realistically depict the behaviour of variability between the two phases. Therefore a series of dividing lines, starting with 1967–8 as the beginning of the second phase, are considered. The least squares regression estimates of the trend lines for the different phases based on the new dividing lines are presented in Table 9.3.

A comparison of the behaviour of absolute production variability (standard deviation) between the two phases clearly indicates that whereas the first period is characterized by an increasing tendency, the second period shows the opposite. The statistical quality of the estimates, however, is not very good, especially in terms of explanatory power. Relatively speaking, the statistical quality of the estimates is much better in terms of explanatory power and statistical significance in the first period compared to that of the second. A similar behavioural pattern can be observed in case of relative production variability. During the first phase it shows a *weak* rising trend while a *relatively* stronger tendency to fall is observed during the second. The explanatory power and statistical significance of the coefficients seem to improve as the dividing line is shifted forward.

An examination of the behaviour of absolute and relative yield variability indicates much the same pattern as that of foodgrain production. In the first period, standard deviation of yield shows a *weak* rising tendency, while a *weak* falling tendency can be observed in the second. The relative variability of yield shows a *weak* declining

Table 9.3 Trends in foodgrain production and yield variability (standard deviation and coefficient of variation) for different phases, based on technological considerations: Bangladesh, 1949–50 to 1981–2

Period	Intercept	Coefficient	R^2	t-value
Standard deviation: Production				
		Phase 1		
1949–66	3.90848	0.1976	0.3352	2.84[a]
1949–67	3.88921	0.2010	0.3801	3.23[a]
1949–68	3.92414	0.1951	0.4020	3.48[a]
1949–69	4.08182	0.1702	0.3542	3.23[a]
		Phase 2		
1967–82	7.93177	−0.1240	0.1478	1.56[c]
1968–82	7.87492	−0.1312	0.1382	1.44[c]
1969–82	7.90800	−0.1501	0.1476	1.44[c]
1970–82	8.37264	−0.2270	0.2790	2.06[b]
Coefficient of variation: Production				
		Phase 1		
1949–66	7.38922	0.1287	0.0615	1.02[d]
1949–67	7.45163	0.1177	0.0604	1.04[d]
1949–68	7.55777	0.1000	0.0508	0.98[d]
1949–69	7.78560	0.0641	0.0236	0.68[d]
		Phase 2		
1967–82	9.55162	−0.2840	0.4367	3.29[a]
1968–82	9.41238	−0.2995	0.4179	3.05[a]
1969–82	9.38783	−0.3312	0.4250	2.98[a]
1970–82	9.85643	−0.4312	0.5784	3.88[a]
Standard deviation: Yield				
		Phase 1		
1949–66	4.28170	0.0857	0.0789	1.17[d]
1949–67	4.44679	0.0566	0.0396	0.55[d]
1949–68	4.58071	0.0343	0.0166	0.84[d]
1949–69	4.72519	0.0115	0.0021	0.20[d]
		Phase 2		
1967–82	4.82926	−0.0205	0.0084	0.34[d]
1968–82	5.03525	−0.0448	0.0341	0.68[d]
1969–82	5.29772	−0.0802	0.0924	1.11[d]
1970–82	5.76758	−0.1490	0.2863	2.10[b]

continued on p. 196

Table 9.3   continued

Period	Intercept	Coefficient	R^2	t-value
Coefficient of variation: Yield				
		Phase 1		
1949–66	6.12158	0.0264	0.0045	0.27[d]
1949–67	6.31775	–0.0082	0.0005	0.09[d]
1949–68	6.47057	–0.0336	0.0092	0.41[d]
1949–69	6.62892	–0.0586	0.0305	0.77[d]
		Phase 2		
1967–82	5.53114	–0.0891	0.1137	1.34[d]
1968–82	5.74233	–0.1213	0.1752	1.66[c]
1969–82	6.01043	–0.1662	0.2706	2.11[b]
1970–82	6.50207	–0.2484	0.5102	3.38[a]

[a] Significant at 1 per cent level;
[b] Significant at 5 per cent level;
[c] Significant at 10 per cent level;
[d] Not significant.

tendency in the first period and a *relatively stronger* tendency to fall in the second. The statistical quality of the estimates seem to show gradual improvement with a forward shift in the dividing line.

A comparison of the estimates set out in Table 9.2 with those of Table 9.3 indicates that the rise in variability during the 1949–60 period is much stronger than in the case of the redefined first phases. By the same token, the fall seems to be stronger in the newly defined second phases compared to the 1961–82 period. Thus, irrespective of how one draws the dividing line between the two phases, that is, on observed mathematical grounds or on *a priori* considerations, the period associated with the new technology can in no way be identified with a period of rising variability in foodgrain production and yield. If anything, with the forward shift in the time cut-off points, both measures of variability show an increasingly stronger tendency to fall, that is, as the Green Revolution becomes more firmly established.

9.5   REASONS FOR THE OBSERVED BEHAVIOUR OF VARIABILITY

From this chapter it can be seen that variability of foodgrain production and yield tended to increase in Bangladesh up to the early 1960s.

After this period, there is evidence of a fall in variability, and in some cases a downward trend in variability has become apparent. These changes have been associated with the introduction of the new agricultural technology.

The present analysis does not support Hazell's (1982, p.10) earlier contention that increased yield instability inevitably accompanies rapid growth in agricultural production. In this context, Ray (1983, p.476) rightly points out that, 'certainly with rapid growth, stability can be achieved if the environment for crop production is brought under human control. Even with a slower growth rate, production can be made more unstable by changing prices and other controllable factors'. As demonstrated earlier in Chapter 2, the growth rates in both production and yield of foodgrains in Bangladesh are higher during the second period than the first. Yet there is no evidence of increased instability in either production or yield in the second period. One can see from Hazell (1985, p.150) that the coefficient of variation of rice production is South Asia which includes India, Bangladesh, Bhutan, Burma, Nepal and Sri Lanka (Hazell, 1985, p.147) has declined by over 36 per cent between the two periods. Of these five countries Bangladesh alone produced nearly 54 per cent of rice in the region during the later period.[4] Given this relative weight of Bangladesh's share in the total production of rice in South Asia during that period, one might attribute some of the fall in variability to that in Bangladeshi rice production. Given the overwhelming importance of rice in total foodgrains in Bangladesh, this may have led to a reduction in the foodgrain production variability during the period 1971–2 to 1982–3.[5] This seems to be consistent with the findings of the present study.

Is it possible to be more specific about the factors associated with or responsible for the change in trend, if not the reversal of trend, in variability of foodgrain production and yield in the second period as compared with the first? This may be possible by considering the sources of growth in foodgrain production in Bangladesh during the two periods. These have been discussed in greater detail in Chapter 2 but may be mentioned briefly in order to understand the causal connection between changes in growth sources and variability in foodgrain production and yield.

In the period prior to the early 1960s, marginal land was brought under cultivation owing to increasing population pressure on existing cultivated land. Net cultivated area increased from about 8.0 million ha in the late 1940s to about 8.5 million ha in the early 1960s (EPBS, 1969, p.41). During the first period the cropping intensity remained stagnant at about 130 per cent. Foodgrain yield showed little tendency to increase and remained stagnant at around 920 kg per ha. The foodgrain sector languished with a slow rate of growth, and the primary source of this growth was the increase in net cultivated area. The upward trend in net cultivated area continued for a few years into the second period, when it reached a peak of 8.8 million ha during the late 1960s. Throughout the entire period of the 1960s, cropping intensity increased slowly but steadily. During the later part of the second period, the net cropped area declined, and by the first half of the 1980s it fell to around 8.6 million ha (for further details, see Alauddin and Tisdell, 1988b).

The increased instability during the period up to the early 1960s may be attributed in part to marginal land (of inferior quality) being brought under cultivation. The then available technology for foodgrain production did little to augment or preserve the fertility of the soil. However, why did variability not continue to increase during the 1960s even though some virgin land was still being brought under cultivation during that period? This may be the result of three factors at work:

1. After new land is brought in for cultivation its quality seems to attain a certain degree of stability after a few years and farmers become more familiar with its qualities (a 'learning by doing' factor). Furthermore, very risky marginal lands may have been withdrawn from production once their poor qualities became apparent. After the 1960s this seems to have occurred. It is possible, therefore, that the new land that was brought in during the 1950s might have ceased to add to the variability during the second period any more than it did during the first.
2. The introduction of modern agricultural inputs like chemical fertilizers and irrigation at the beginning of the second period is likely to have to some extent helped maintain the fertility of existing land under cultivation. This may have compensated for any increased variability that may have resulted from the extension of the frontiers of cultivated land during the earlier part of the second period.
3. During the later part of the 1960s biological innovations like HYVs of rice and wheat were introduced. This made it possible to replace traditional rice cultivation by a whole range of new technologies. This had important implications in terms of higher productivity per unit area as well as greater choice and flexibility. The new technologies enabled the incidence of multiple cropping

to increase. The growth of foodgrain production was more rapid during the second period compared to that of the first, and the dominant source of this growth was increase in yield per hectare (Alauddin and Tisdell, 1986c; Chapter 2). Increased frequency of cropping during a year following the introduction of HYVs and related technologies was a major factor in yield increase.

An increase in the incidence of multiple cropping can reduce the overall (annual) variability of production and *a fortiori* reduce the coefficient of variation of production. In Bangladesh the introduction of the new technology has permitted the use of land for different crops of foodgrains over the whole year, e.g. its use for *kharif* (summer and wet season) foodgrains (*aus* and *aman* rice) and *rabi* (dry season) foodgrains (*boro* rice and wheat). Furthermore, within one season, land can be allocated between local and HYVs of the same foodgrain crop, e.g. local and HYVs of *aus* rice. When technological innovations reduce constraints on the use of a non-renewable resource like land and allow the cultivation of two or more crops instead of only one crop during a year on the same plot of land, or different varieties of the same crop during the same crop season, the chance of a total crop failure considered annually can be reduced. The situation is akin to diversification of portfolios as a hedge against uncertainty (cf. Markowitz, 1959; Anderson *et al.*, 1977).

Consider now the impact multiple cropping might have had on the variability of yield. A casual observation of the relevant columns in Table 9.4 indicates that overall there is a tendency for the relative yield variability to decline with increase in the intensity of cropping. The relationship seems to be stronger towards the later part of the series. These aspects can be better illustrated by Figure 9.5 where the coefficient of variation of yield against intensity of cropping is plotted. The impact of multiple cropping on foodgrain yield variability in different phases comes into clearer focus if one considers the estimated regression equations involving the two variables. The statistical quality of the estimates pertaining to the second phase is consistently better than that of the earlier or the entire period. Furthermore, it seems to improve within the first phase as later year cut-off figures are used in regression estimates. During the second phase significant improvement in the statistical quality of the estimates takes place when the cut-off point is shifted forward to include observations corresponding to the period where some progress has already been made with regard to the expansion of the HYV-related technology.

Table 9.4   Impact of intensity of cropping on coefficient of variation of yield in different phases based on Plot 5 and technological considerations: Bangladesh, 1949–1982

Period	Intercept	Coefficient	R^2	t-value
		Phase 1		
1949–60	95.69082	–0.6876	0.1107	1.12[d]
1949–66	46.21385	–0.3040	0.2464	2.29[b]
1949–67	39.27908	–0.2507	0.2622	2.46[b]
1949–68	35.53937	–0.2220	0.2809	2.65[a]
1949–69	34.80203	–0.2164	0.3210	3.00[a]
		Phase 2		
1961–82	20.82291	–0.1108	0.4060	3.70[a]
1967–82	31.67769	–0.1837	0.5807	4.40[a]
1968–82	33.10051	–0.1929	0.6288	4.69[a]
1969–82	34.10286	–0.1993	0.6687	4.92[a]
1970–82	35.73924	–0.2096	0.7681	6.04[a]
Entire period	23.61529	–0.1301	0.3576	4.22[a]

[a] Significant at 1 per cent level;
[b] Significant at 5 per cent level;
[d] Not significant.

Cropping intensity may be used as a rough proxy for technological change in Bangladeshi agriculture. Irrigation, fertilizer and HYVs all tend to increase multiple cropping. Because of complementarity between these inputs, one would expect a higher degree of multi-collinearity between them. In addition, the 'learning by doing' factor would contribute to a reduction in relative variability of production and yields as experience with the new technology proceeds.

A number of factors, apart from those suggested elsewhere in this chapter, may have contributed to the declining relative variability of yield after the introduction of HYVs. In the beginning, experiment stations often test and release a wide range of varieties. Some of these prove to have higher variability under field conditions than is apparent under experimental conditions, and are discontinued. Others may initially be applied outside the regions ecologically most suited for them. Thus general learning about the ecological suitability and appropriateness of introduced varieties to particular areas takes place over time. *In addition*, individual farmers become more familiar with the environmental and husbandry requirements of new varieties so they can improve their cultural practices. This is an individual 'learning by experience' phenomnenon. Both of these

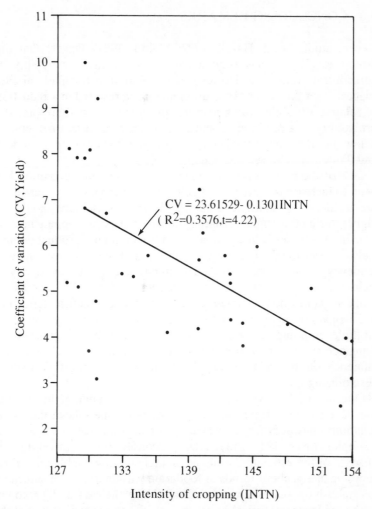

Figure 9.5   Coefficient of variation of foodgrain yield in relation to intensity of cropping, Bangladesh, 1949–50 to 1982–3

experimental factors will tend to reduce yield variability with the passage of time.

The findings that emerge from the discussion surrounding Figure 9.5 and Table 9.4 are consistent with those in the preceding section. Thus the available evidence does not support the hypothesis that the Green Revolution has led to any increase in the variability in food-grain production and yield, at least in the case of Bangladesh. Indeed, there is some evidence to the contrary.[6]

9.6   CONCLUDING REMARKS

Recent studies (e.g. Hazell, 1982; Mehra, 1981) suggest that the Green Revolution has been a source of increased variability of agricultural production. However, examination of Bangladeshi data suggests that the Green Revolution may have resulted in a reduction of relative variability of agricultural production and, therefore, the probability of a fall in yield and production of a certain percentage below the trend may have been reduced (see, for example, Alauddin and Tisdell, 1988d; see also Chapter 10).

One of the most important impacts of the Green Revolution has been to increase the intensity of cultivation or cropping by increasing the incidence of multiple cropping. Even if variability should be higher during the original cropping period, the extra cropping in each year in most cases is likely to add stability in annual production and yield. Even though absolute variability might in some cases show a tendency to increase, increased (annual) average production and yield can bring about a decrease in relative variability. Increased usage of controlled complementary inputs like chemical fertilizers and irrigation might also have reinforced the moderating impact of multiple cropping. Because these factors have been significant in Bangladesh, the Green Revolution appears to have had a stabilizing influence on the relative variability of production rather than a destabilizing one.

One needs, however, to qualify the present findings in two respects. First, the present analysis is based on the official data, the reliability and accuracy of which is not beyond question (see, for example, Boyce, 1985; Pray, 1980). Secondly, the statistical quality of some of the estimates, while adequately casting doubts on earlier views or theses about trends in foodgrain variability, is not sufficient to establish the opposite view or thesis. Nevertheless (taking account of the evidence, statistical or *otherwise*), the thesis cannot be lightly dismissed that the Green Revolution has reduced, and is continuing to reduce, as it proceeds, the relative variability of foodgrain production and yield.

10 New Agricultural Technology and Instability of Foodgrain Production and Yield: An Inter-Temporal and Inter-Regional Perspective

10.1 INTRODUCTION

Chapter 9 identified limitations of the methodological framework of Hazell (1982; 1984; 1985) and Mehra (1981) in investigating variability of Indian foodgrain production and yield, and examined the trends in absolute and relative variability in Bangladeshi foodgrain production and yield using an alternative methodology. In this chapter evidence from Bangladesh about changes in variability of production and yield of foodgrains prior to and following the introduction of the HYVs of cereals and associated techniques is outlined and investigated at a more disaggregated level. First of all, the methodology employed for analysing the data is specified. Trends in foodgrain production and yield variability for Bangladesh as a whole, as well as for the main regions (districts) of Bangladesh, are then discussed. District (regional) data are also used to examine how variability of foodgrain production and yield have altered with the introduction of the new agricultural technology. Because rates of adoption and spread of the new technology by regions are uneven (see, for example, Alauddin and Mujeri, 1986a); Alauddin and Tisdell, 1988d; 1988e), this chapter investigates whether variability in foodgrain production and yield variability are systematically related to the rate of diffusion of the new agricultural techniques.

More specifically, this chapter seeks to find answers to the following questions. To what extent have these developments led to alterations in the variability of overall foodgrain production and yield? Is there any significant difference in production and yield variability between traditional and modern foodgrain varieties? Has irrigation had a stabilizing impact on production and yield fluctuations? All these questions are worth considering in the Bangladeshi context. One of the major findings that emerges from the Mehra (1981) study is the claim that irrigation reduces yield instability. An analysis of Mehra's data (p.29) for fifteen Indian States supports her claim. Coefficients of correlation between the ranks of both standard deviation and coefficient of variation of yield on the one hand and the percentage area irrigated were 0.1096 and −0.4714 respectively. Further analysis of Mehra's data indicates that the standard deviation of yield has a tendency to increase with per cent area under HYV and fertilizer application (rank correlation coefficients respectively are 0.4688 and 0.4528) and decline with higher intensity of cropping (rank correlation coefficient = −0.0783). Coefficient of variation of yield shows a tendency to decline with increased use of modern technology. The coefficients of rank correlation with per cent area under HYV, fertilizer application and cropping intensity respectively are −0.1500, −0.1179 and −0.5584. This is consistent with the hypothesis that modern technology has tended to reduce overall yield variability rather than increase it, a matter which will now be specifically explored using Bangladeshi national and regional data.

10.2 METHODOLOGICAL FRAMEWORK OF ANALYSIS

Linear trend lines[1] were fitted to relevant Bangladeshi data (foodgrain production and yield) for two periods: 1947–8 to 1968–9 (period of traditional technology) and 1969–70 to 1984–5 (period of modern technology) for Bangladesh as a whole and for the districts or regions (same as the ones mentioned in Chapter 4 and shown in Figure 4.1). It should be noted that national data on all HYVs are available since 1967–8, but district level data on all HYVs are available only from 1969–70. To provide a common basis of comparison between national and regional results, 1969–70 has been taken as the dividing line. As there was little penetration of the new technology in 1967–8 and 1968–9, the time cut-off point adopted in this chapter is unlikely to prejudice overall findings. Variability of production and yield is measured in terms of their yearly deviations from the estimated trend values, that is using a similar approach to Hazell (1982; 1985).

Let $X_t = a + bt + u_t$ (10.1)

where x_t denotes the dependent variable (production or yield), t is time and u_t is the random disturbance term with zero mean and variance s^2. Then absolute variation around the trend can be measured by the standard error of estimate (*SE*) of the relevant variable which is given as

$$SE = ((X_t - \hat{X}_t)^2/(n - 2))^{1/2}$$ (10.2)

where $\hat{X}_t$ is the estimated trend value of X_t and n is the number of observations. Variance is defined as the square of standard error of estimate. The relative variability, the coefficient of variation[2] (*CV*), is then defined as

$$CV = (SE/\bar{X}) \times 100$$ (10.3)

$\bar{X}$ being the mean of X_t.

Since the objective of this chapter is principally to consider inter-district variability in foodgrain production relative variability is more appropriately measured by the coefficient of variation than the standard error of the estimate or the unexplained variation. Because different districts have different mean levels of production and yield (see Table 10.3), standard deviations are likely to be higher for the districts with higher absolute values of those variables. A comparison of inter-district standard deviations, therefore, would be of little value. It is, therefore, worth while to concentrate on an inter-district analysis of relative variability.

One way of examining the implications of production and yield variability is to calculate probability of these variables falling 5 per cent or more below the trend for each year for each district. One difficulty with such an exercise is the unknown distribution of possible production and yield outcomes because of a single observation for each variable in one year. Since the variance of production and yield around the trend can be estimated for the entire period, it is possible to obtain average probabilities on the assumption of a constant variance for all years during the period. These probabilities can be obtained as follows (see Hazell, 1985, p.149).

Let $\hat{X}_{it} = \bar{X}_i + u_t$, where $\bar{X}_i$ is the overall mean for the ith district, $\hat{X}_{it}$ is the trend value of X for the ith district in year t within a period, and u_t is the deviation from the mean in year t. Then the probability of a shortfall of 5 per cent or more below the trend can be estimated from $Pr(u_t/SE < -0.05\bar{X}/SE)$ where SE is the standard error of estimate. Assuming u_t is approximately normally distributed the derived probability can be obtained from the tabulated values of cumulative normal distribution.

While this approach, adopted by Hazell (1982; 1985), is useful in making a comparative analysis of variability across districts, one must take into account the limitations identified in the preceding chapter. The methodology used here, which is similar to that used by Hazell (1982; 1985), does not provide information about the changing behaviour of variability with the passage of time. Nevertheless, the approach of Hazell (1982) is adopted for two reasons: First, this method has not been previously applied to Bangladeshi data. It has been applied to Indian data and it is useful to have a comparison between India and Bangladesh. Secondly, the objective is to examine the behaviour of variability between districts. This chapter first considers Bangladeshi national time series data and then focuses attention on regional data looking at trends over time and inter-regional (cross-sectional) differences.

10.3   OBSERVED BEHAVIOUR OF VARIABILITY 1947–8 TO 1984–5

Variability of foodgrain yield and production for Bangladesh as a whole can now be examined. Table 10.1 sets out average production and yield of various food crops for Bangladesh as a whole for 1947–8 to 1968–9 (Period 1) and 1969–70 to 1984–5 (Period 2). In the first period there was little penetration of HYV technology, whereas the second period corresponds to significant adoption of HYV technology. Table 10.1 also sets out the respective coefficients of variation and the probabilities of a 5 per cent (or greater) fall in yield and production below the trend. Inter-temporal comparison shows significant rises in production and yield of all varieties of foodgrains. While rice production rose significantly, wheat production registered a phenomenal increase between the two periods. Foodgrain yield increased by 25 per cent. The real increase in yield would be higher when one takes into account the increase in cropping intensity from 130 per cent to 146 per cent over the same period (see Alauddin and Tisdell, 1986c).

Furthermore, Table 10.1 indicates a substantial decline in relative

Table 10.1 Changes in average quantities, coefficients of variation and probability of a 5 per cent or greater fall below the trend: Foodgrain production and yield, Bangladesh, 1947–8 to 1968–9 (Period 1) and 1969–70 to 1984–5 (Period 2)

(a) Production

Crop	MV (00 m/ton)			CV (%)			PROB5 (%)		
	Period 1	Period 2	% Change	Period 1	Period 2	% Change	Period 1	Period 2	% Change
Rice	8674	12615	45.43	9.15	6.11	-33.23	29.26	20.67	-29.36
aus	2150	2982	38.70	10.74	9.49	-11.67	32.10	29.91	-6.82
aman	6014	7126	18.49	10.36	8.18	-21.02	31.46	27.06	-13.99
boro	509	2507	392.53	44.01	15.04	-65.83	45.46	36.99	-18.63
Wheat	34	544	1500.00	41.18	33.82	-17.86	45.18	44.12	-2.35
Kharif	8164	10108	23.81	9.08	7.93	-12.58	29.09	26.43	-9.14
Rabi	544	3051	460.85	43.57	16.49	-62.16	45.42	38.10	-16.12
All food	8708	13159	51.11	9.16	6.18	-32.58	29.26	20.93	-28.47

continued on page 208

Table 10.1 *continued*

(b) Yield

Crop	MV (kg/ha)			CV (%)			PROB5 (%)		
	Period 1	Period 2	% Change	Period 1	Period 2	% Change	Period 1	Period 2	% Change
Rice	1005	1242	23.58	6.96	4.03	−42.20	23.64	10.71	−54.69
aus	863	945	9.50	8.23	6.14	−25.40	27.12	20.76	−23.45
aman	1049	1217	16.01	8.48	6.24	−26.39	27.79	21.16	−23.86
boro	1191	2179	82.96	15.28	7.66	−49.85	37.18	25.72	−30.82
Wheat	622	1542	147.91	13.18	13.29	0.84	35.24	35.35	0.31
Kharif	994	1122	12.88	7.44	5.44	−26.97	25.08	17.88	−28.71
Rabi	1125	2083	85.16	15.20	6.52	−57.05	37.11	22.18	−40.23
All food	1003	1256	25.22	6.98	4.06	−41.82	23.70	10.91	−53.97

Notes: MV refers to mean values, CV is coefficient of variation, while PROB5 represents probability of a 5 per cent or greater fall below the trend. *Kharif* refers to summer and rainy season rice crops, i.e. *aus* and *aman* rice while *rabi* refers to dry season food crops, i.e. *boro* rice and wheat. All food is the total rice and wheat crops.

Source: Adapted from BBS (1976, pp.1–2, 4–10, 12–5, 26–9; 1979, pp.168–71; 1980a, pp.20–5, 30–1, 33–4, 36–7, 46–52; 1982, pp.232, 235–8, 240–1; 1984a, pp.39, 42; 1984b, pp.249–52, 255; 1985a, pp.32–72; 1985b, pp.301, 310; 1986a, pp.39, 47–50, 53); BRRI (1977, p.89); World Bank (1982, Tables 2.5–6). Local *aman* includes transplant and broadcast *aman*. High-yielding varieties of *aus*, *aman* and *boro* rice also include *Paijam* varieties.

variability of foodgrain production and yield between the two periods. Also for all foodgrain crops, except wheat, the probabilities of a 5 per cent or greater fall in production and yield below the trend have fallen markedly. In general, irrigated food crops show greater declines in variability compared with rainfed ones. This is confirmed by the fact that reductions in the coefficients of variation have been much greater for *rabi* foodgrains (*boro* rice and wheat taken together) than for *kharif* foodgrains (*aus* and *aman* rice taken together). This suggests that irrigation has reduced instability. It is interesting to note that the production and yield of wheat became more variable during the second period. This is probably in part because only a little over a third of the area under wheat cultivation is irrigated compared to about 90 per cent for *boro* rice. Therefore, the stabilizing impact of irrigation is likely to be much less prominent for wheat than it is for *boro* rice. The behaviour of overall variability in foodgrain production and yield presents a contrasting picture to that for Indian agriculture (cf. Mehra, 1981, p.18).

One can employ the Hazell variance decomposition method (Hazell, 1982) to identify the apparent sources of change in production variance following the introduction of the new agricultural technology. While there are some reservations about this method (see Alauddin and Tisdell, 1986c) because of its similarity with that of Minhas and Vaidyanathan (1965), the method has been widely used despite its mechanistic nature. The Hazell decomposition method was applied, using Table 10.1 in order to assess the relative importance of various components of change (Hazell, 1982, Table 6, p.20 and Equation 13, p.21). However, the results were statistically insignificant and indicate that overall production variance only increased by 4 per cent between the periods analysed. Furthermore, the change is numerically too small to enable a robust analysis of components of change to be completed.

10.4 INTER-REGIONAL AND INTER-TEMPORAL ANALYSIS OF VARIABILITY

In this section, production and yield variability of all foodgrains are examined for each district and then compared between time periods. Secondly, variability of foodgrain yield and production are examined within districts and compared for different seasons and varieties of crops and between districts.

10.4.1   Inter-temporal Comparison of District-Level Variability

Table 10.2 sets out selected characteristics of foodgrain production and yield on a regional basis prior to and following the Green Revolution. It shows that alterations in the overall foodgrain production and yield following the Green Revolution in Bangladesh vary considerably between districts. The districts of Chittagong, Noakhali, Mymensingh, Rangpur, Bogra and Dinajpur have experienced large production increases well in excess of the national increase (of 51 per cent), whereas there was virtually no increase in average foodgrain production for Faridpur and Khulna. A similar picture emerges concerning average yield. Chittagong, Noakhali, Comilla, Bogra, Mymensingh and Dhaka registered significant yield increases over the national average of 25 per cent. There was very little yield increase in Faridpur, Barisal, Khulna and Jessore, while moderate increases were recorded in the remaining seven districts.

As for production and yield variability, significant declines occurred in most districts and, in fact, production variability rose in only Comilla and Sylhet. Relative yield variability declined by over 30 per cent in 12 districts and only in Sylhet had it slightly increased. The probability of production falling 5 per cent or more below trend decreased in all but two of the 17 districts. It decreased by more than 28 per cent (the national average decline) in Chittagong, Dhaka, Mymensingh and Rangpur while all other districts recorded moderate to insignificant declines. The probability of annual yield falling 5 per cent or more below trend declined in all districts. Four districts, namely Mymensingh, Dinajpur, Chittagong and Dhaka, experienced declines of more than the average national decline (of 55 per cent). Significant declines were also recorded in Bogra, Pabna, Rangpur and Rajshahi. Overall there have been substantial declines in the probability of yield falling 5 per cent or more below the trend. On the whole, the regional data support the hypothesis that a reduction in overall yield variability has coincided with the introduction of the new technology.

10.4.2   Behaviour of Variability and New Technology

As can be seen from Table 10.2, significant inter-district differences exist in the variability of foodgrain production, ranging from 5.64 per cent for Mymensingh to 16.08 per cent for Sylhet. Faridpur, Chittagong Hill Tracts, Dinajpur, Pabna, Comilla, Khulna and Kushtia

Table 10.2 Changes in average quantities, coefficients of variation and probability of a 5 per cent or greater fall below the trend: Foodgrain production and yield, Bangladesh districts, 1947–8 to 1968–9 (Period 1) and 1969–70 to 1984–5 (Period 2)

(a) Production

District	MV (00 m/ton) Period 1	Period 2	% Change	CV (%) Period 1	Period 2	% Change	PROB5 (%) Period 1	Period 2	% Change
Dhaka	508.0	741.1	45.89	17.30	5.96	−65.53	38.63	20.11	−47.94
Mymensingh	1252.2	2215.8	76.95	11.42	5.64	−50.60	33.07	18.72	−43.39
Faridpur	466.4	479.3	2.77	17.88	13.50	−24.51	38.97	35.57	−8.72
Chittagong	400.3	727.5	81.74	17.24	6.19	−64.11	38.59	20.96	−45.68
Chittagong HT	71.8	101.5	41.36	18.80	13.89	−26.12	39.51	35.94	−9.04
Noakhali	358.5	639.5	78.38	11.99	11.27	−6.00	33.83	32.89	−2.78
Comilla	648.0	963.5	48.69	10.23	11.81	15.44	31.25	33.61	7.55
Sylhet	770.2	1101.6	43.03	13.40	16.08	19.98	35.46	37.82	6.65
Rajshahi	575.1	808.5	40.58	18.62	10.75	−42.28	39.44	32.10	−18.61
Dinajpur	415.8	684.7	64.67	20.42	12.36	−39.49	40.33	34.28	−15.00
Bogra	317.1	544.4	71.68	19.24	9.53	−50.44	39.74	30.01	−24.48
Rangpur	697.6	1221.6	75.11	13.19	7.47	−43.33	35.24	25.17	−28.57
Pabna	284.8	419.6	47.33	16.40	12.65	−22.82	38.02	34.65	−8.86
Khulna	492.8	531.3	7.81	25.18	13.68	−45.66	42.11	35.75	−15.10
Kushtia	195.9	264.8	35.17	16.23	12.73	−21.60	37.91	34.72	−8.41
Jessore	413.7	572.4	38.36	13.78	8.07	−41.42	35.95	26.79	−25.48
Barisal	813.3	1040.0	27.87	16.98	9.20	−45.81	38.44	29.36	−23.62

continued on page 212

Table 10.2 continued

(b) Yield

District	MV (kg/ha)			CV (%)			PROB5 (%)		
	Period 1	*Period 2*	*% Change*	*Period 1*	*Period 2*	*% Change*	*Period 1*	*Period 2*	*% Change*
Dhaka	1026	1334	30.02	10.92	4.87	−55.36	32.71	15.25	−53.38
Mymensingh	1005	1317	31.04	8.26	3.64	−55.87	27.26	8.50	−68.82
Faridpur	833	874	4.92	13.32	10.18	−23.58	35.39	31.17	−11.92
Chittagong	1141	1824	59.86	13.41	4.82	−64.02	35.46	15.01	−57.67
Chittagong HT	1147	1463	27.55	12.99	9.57	−26.33	35.05	30.05	−14.26
Noakhali	953	1297	36.10	11.75	9.48	−19.31	33.54	29.91	−10.82
Comilla	1023	1392	36.07	9.09	7.11	−21.77	29.12	24.11	−17.20
Sylhet	1077	1316	22.19	7.61	7.45	−2.19	25.56	25.11	−1.76
Rajshahi	952	1167	22.58	12.71	7.71	−39.32	34.72	26.85	−22.67
Dinajpur	1008	1237	22.72	13.39	5.01	−62.58	35.46	14.19	−59.98
Bogra	1024	1362	33.01	14.84	6.90	−53.50	36.80	23.46	−36.25
Rangpur	967	1253	29.58	10.34	6.70	−35.17	31.39	22.78	−27.43
Pabna	875	1040	18.86	14.74	7.40	−49.78	36.73	24.99	−31.96
Khulna	1084	1172	8.12	16.88	10.84	−35.81	38.36	32.24	−15.95
Kushtia	884	1123	27.04	14.82	9.71	−34.50	36.80	30.68	−16.63
Jessore	969	1094	12.90	11.87	8.13	−31.45	33.40	27.26	−18.38
Barisal	1044	1125	7.76	13.70	9.16	−33.16	35.76	29.26	−18.18

Source: Based on data from sources mentioned in Table 10.1

have production variabilities towards the upper limit of this range. Turning to the question of overall yield variability (between districts) one can see that it ranges between 3.64 per cent in Mymensingh to 10.84 per cent in Khulna. Chittagong, Dhaka, Dinajpur, Comilla, Bogra and Rangpur have relative yield variabilities towards the lower bounds of the range. Significant inter-regional differences in the variability of seasonal and varietal yields can be observed.

Table 10.3 also sets out average production and yield of foodgrains by season and variety for the 17 districts for the period 1969–70 to 1984–5. Also presented are the corresponding coefficients of variation. Significant inter-district variations in overall yield are noticeable, ranging from 874 kg in Faridpur to 1824 kg in Chittagong. Seasonal yield differences also exist. Chittagong has the highest yields during the *kharif* season (1607 kg) as well as during the *rabi* season (2652 kg). Faridpur has the lowest yield during the *kharif* season (733 kg) while Sylhet has the lowest yield during the *rabi* season (1630 kg). Faridpur, Khulna, Kushtia, Pabna, Rangpur, Rajshahi, Mymensingh, Jessore and Dinajpur have lower *rabi* season yields than the remaining seven districts.

Table 10.3 indicates that the aggregate regional production of HYV foodgrains is relatively more variable than that of the local varieties, with the exception of the districts of Chittagong, Comilla, Kushtia, and Chittagong Hill Tracts, where HYV production is relatively less variable than local variety foodgrain production. *Rabi* foodgrain production is relatively more variable than *kharif* for all districts except Chittagong, Comilla, Rajshahi, Pabna, Khulna and Barisal, where the opposite is the case. When compared with the relevant column in Table 10.2, the overall annual relative variability of foodgrain production and yield is seen to be generally smaller than that of foodgrains in different seasons (*rabi* and *kharif*) and of different varieties (local and HYV).

Similar comparative variations can be observed for varietal and seasonal production variabilities. In general Dhaka, Mymensingh, Chittagong and Rajshahi show relatively lower values of variability. HYVs show highest variability in Dinajpur, closely followed by Faridpur, Rangpur, Pabna and Sylhet, and lowest in Dhaka, followed by Chittagong, Mymensingh, Jessore and Comilla.

Table 10.3 also sets out the relative yield variability of foodgrains for different varieties and seasons. In the majority of the districts, foodgrain yields during the *kharif* season are less variable than those during the *rabi* season. However, for seven districts, namely,

Table 10.3 Foodgrain production and yield: Mean values and coefficients of variation by season and variety, Bangladesh districts, 1969–70 to 1984–5

(a) Production

District	Mean production ('000 m tons)				Coefficient of variation (%)			
	Rabi	Kharif	Local	HYV	Rabi	Kharif	Local	HYV
Dhaka	262.0	479.0	429.3	311.7	11.41	8.22	8.80	9.98
Mymensingh	670.5	1545.3	1325.2	890.6	21.68	7.78	7.33	9.72
Faridpur	118.3	361.1	372.1	107.2	39.39	15.54	14.83	40.76
Chittagong	230.0	497.6	254.6	473.0	8.74	9.65	13.31	10.02
Chittagong HT	22.2	79.4	42.6	59.0	22.07	14.86	30.99	13.73
Noakhali	131.0	508.5	354.9	284.6	21.68	14.24	13.89	19.29
Comilla	289.6	673.9	515.9	447.6	12.74	15.39	17.56	14.23
Sylhet	404.9	696.8	847.0	254.6	23.91	14.55	20.87	30.09
Rajshahi	157.2	651.3	634.6	173.9	20.48	11.45	12.72	24.67
Dinajpur	70.8	613.9	522.0	162.7	40.25	12.56	15.71	35.34
Bogra	101.0	443.5	344.9	199.5	38.71	6.74	9.94	23.16
Rangpur	139.9	1081.8	918.4	303.3	40.39	5.19	8.79	29.64
Pabna	113.6	306.0	305.5	114.1	38.03	19.38	18.13	33.74
Khulna	43.1	488.2	437.5	93.8	26.91	15.12	16.75	16.95
Kushtia	70.2	194.50	152.3	112.5	19.37	16.91	19.50	15.64
Jessore	75.5	496.9	429.6	142.9	45.03	11.99	12.06	18.47
Barisal	136.2	903.8	793.2	246.8	28.63	13.18	15.62	22.41

(b) Yield

District	Mean yield (kg per hectare)				Coefficient of variation (%)			
	Rabi	Kharif	Local	HYV	Rabi	Kharif	Local	HYV
Dhaka	2370	1077	988	2766	7.89	6.22	5.87	10.63
Mymensingh	2051	1140	1020	2574	10.82	7.28	4.90	10.10
Faridpur	1981	733	732	2915	9.14	9.28	8.88	14.54
Chittagong	2652	1607	1278	2524	10.41	7.72	7.98	12.00
Chittagong HT	2268	1331	956	2343	11.68	10.67	13.81	17.58
Noakhali	2533	1162	1005	2368	11.96	12.82	11.34	14.82
Comilla	2222	1201	1058	2369	8.10	10.16	9.26	13.93
Sylhet	1630	1182	1162	2455	12.76	7.44	9.47	13.77
Rajshahi	1940	1062	1028	2572	7.78	8.00	8.85	13.18
Dinajpur	1923	1192	1100	2258	7.75	5.45	6.09	9.74
Bogra	2226	1254	1119	2402	10.20	6.54	6.70	18.19
Rangpur	1966	1195	1097	2413	10.58	5.69	7.11	12.52
Pabna	1904	883	861	2623	8.30	10.98	10.80	12.35
Khulna	1864	1134	1073	2379	10.62	12.08	12.30	20.34
Kushtia	1852	980	839	2362	21.76	13.47	13.59	15.92
Jessore	2095	1019	932	2571	13.79	6.97	7.19	13.03
Barisal	2459	1038	965	2486	8.09	12.43	13.78	11.75

Source: Based on data from sources mentioned in Table 10.1

Noakhali, Khulna, Faridpur, Comilla, Rajshahi, Pabna and Barisal, *rabi* season yields are less variable than those during the *kharif* season. Local variety yields are relatively less variable than those of the HYVs for all districts except Barisal. Overall yields are less variable than the component yields (cf. Table 10.1). Again this may be due to aggregation and lack of perfect correlation between components, with the exception of Barisal. In Rangpur, *kharif* season yields are less variable than the overall yield.

What the preceding discussion seems to indicate is that on a *seasonal* basis HYVs show greater relative variability than local varieties, but not on an *annual* basis. This may be due to averaging and other factors. Thus the fact that HYV yields are more stable than traditional varieties on an annual basis but are more variable within seasons could mean that either (1) HYVs are more stable than traditional varieties in one season but not in the other, or (2) HYVs have more negative correlations between seasons than do traditional varieties. An important contribution of the Green Revolution seems to have been to moderate annual relative production and yield variability. However, it does not appear to have reduced seasonal or varietal variabilities either at all or only to an extent.[3]

To what extent is the observed behaviour of foodgrain production and yield variability between districts related to the introduction of the new technology in Bangladesh? To what extent do cross-sectional data suggest that there is a relationship between the adoption of the new technology and variability of food production and yield? Here this matter is considered, bearing in mind that correlation between factors does not necessarily imply a causal relationship between them.

Indicators of adoption of the new technology by districts are set out in Table 10.4 and some indicators of the 1969–70 scenario are stated. Table 10.4 indicates that cropping intensity (INTN), use of fertilizer and irrigation (FRTHEC and PRFAI) tend to go up as the percentage of foodgrain area under HYV (PRHYVF) increases. These aspects come into sharper focus when one plots various indicators of variability against the surrogates for adoption of the new technology and intensity of cropping. Figure 10.1 plots coefficients of variation of production and yield (CVPROD and CVYIELD) against per cent area under HYV foodgrains. In order to make quantitative comparisons of variability across districts, four regression lines were estimated with per cent area under HYV foodgrains as the explanatory

Table 10.4   Selected indicators of spread of the new technology, 1969–70 to 1984–5, and of the 1969–70 scenario: Bangladesh districts

(a)   Indicators of technological change (1969–70 to 1984–5)

District	Cropping Intensity (%)	% Food area (Irrigated)	% HYV area in food	Fertilizer (kg/ha)
Dhaka	145.3	17.1	21.3	39.3
Mymensingh	170.0	17.2	21.5	22.5
Faridpur	157.1	4.7	7.3	8.0
Chittagong	150.1	22.2	49.3	61.8
Chittagong HT	147.7	13.8	40.5	18.6
Noakhali	152.5	10.1	27.5	21.3
Comilla	162.0	14.6	28.9	45.0
Sylhet	136.9	25.7	13.2	11.4
Rajshahi	128.3	13.2	10.5	20.5
Dinajpur	140.9	6.3	13.4	21.8
Bogra	159.5	10.7	21.0	40.8
Rangpur	181.4	5.5	13.3	13.7
Pabna	153.5	7.5	11.1	20.7
Khulna	124.5	5.5	9.4	10.2
Kushtia	133.7	20.9	20.7	35.2
Jessore	135.7	7.1	11.6	18.5
Barisal	139.8	7.6	11.2	11.6

(b)   1969–70 scenario

District	YLD6970 (kg/ha)	IRRI6970 (per cent)	FERT6970 (kg/ha)	POTKHYV (per cent)
Dhaka	1143	8.1	15.8	25.74
Mymensingh	1152	13.7	8.2	55.53
Faridpur	891	2.3	1.9	11.31
Chittagong	1673	16.3	51.4	95.69
Chittagong HT	1482	14.5	11.7	69.12
Noakhali	1193	3.6	11.6	66.12
Comilla	1313	9.9	17.8	34.61
Sylhet	1422	22.5	6.0	71.25
Rajshahi	1170	15.3	7.0	58.16
Dinajpur	1093	5.4	9.5	99.60
Bogra	1055	5.0	17.3	72.29
Rangpur	1141	2.6	5.3	71.35
Pabna	985	3.0	5.4	28.98
Khulna	1118	3.6	6.6	57.82
Kushtia	817	6.8	10.9	89.57
Jessore	974	2.7	6.0	45.23
Barisal	1176	2.7	5.8	65.06

Notes: Figures in part A are based on the 1969–70 to 1984–5 averages. YLD6970, IRRI6970, FERT6970 respectively are foodgrain yield, per cent of foodgrain are irrigated, and amount of fertilizer (nutrients) per hectare of gross cropped area in 1969–70; POTKHYV is land area potentially suitable for *kharif* HYV cultivation as percentage of mean *kharif* area for the period 1969–70 to 1984–5.

Source: Based on data from sources mentioned in Table 10.1 and BBS (1979, pp.162–7, 212; 1982, pp.206–13; 1984a, pp.31–3; 1986a, pp.39–43, 70–2); Bramer (1974, pp.64–5).

variable and each of the other two as a dependent variable. These are illustrated by Figure 10.1.

Even though the signs seem to indicate a decline in variability with increase in the per cent area under HYV foodgrains, one should note the poor quality of the estimates both in terms of explanatory power and statistical significance of the coefficients. In view of this one needs to be wary. While the evidence does not indicate a strong tendency for variability to decline with per cent area under HYV, it may rule out any upward tendency in variability with the percentage area sown to HYVs.

Figure 10.2 depicts the behaviour of the same measure of variability across districts in relation to the intensity of cropping. The relationship of relative variability of foodgrain production and yield to cropping intensity is similar to the one for relative production and yield variability and per cent area under HYV foodgrains. The estimated regression lines are illustrated in Figure 10.2. The estimated relationships are poor, both in respect of R^2 and t-values. One therefore should not place too much significance on these regression lines. Despite this it could be said with a fair degree of confidence that neither *overall* production nor *overall* yield variability show any tendency to increase with an increase in the intensity of cropping. If anything, there is a tendency for these variabilities to decline with increases in the incidence of multiple cropping.

The above evidence suggests that the new agricultural technology has not caused increased variability of foodgrain production and yield on a regional and annual basis. If anything, the new agricultural technology has been an ameliorating influence on such variability.[4]

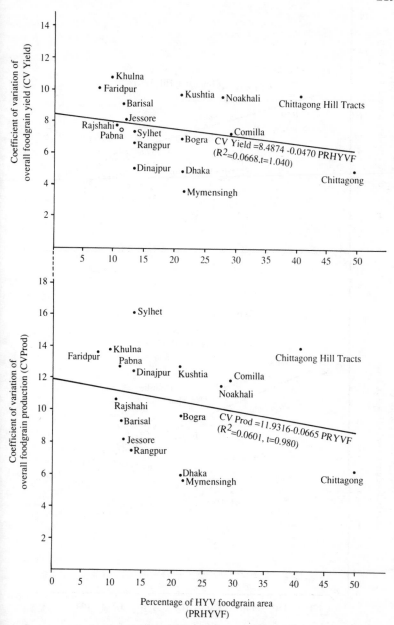

Figure 10.1   *Top*: Coefficient of variation of overall foodgrain yield (CVYIELD) vs percentage of HYV foodgrain area (PRHYVF) for Bangladesh districts, 1969–70 to 1984–5. *Bottom*: Coefficient of variation of overall foodgrain production (CVPROD) vs percentage of HYV foodgrain area (PRHYVF) for Bangladesh districts, 1969–70 to 1984–5

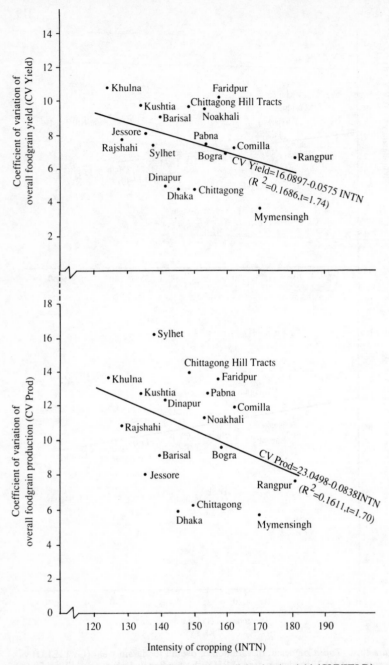

Figure 10.2   *Top*: Coefficient of variation of overall foodgrain yield (CVYIELD) vs intensity of cropping (INTN) for Bangladesh districts, 1969–70 to 1984–5. *Bottom*: Coefficient of variation of overall foodgrain production (CVPROD) vs intensity of cropping (INTN) for Bangladesh districts, 1969–70 to 1984–5

10.5 OBSERVED BEHAVIOUR OF VARIABILITY: UNDERLYING FACTORS

The present analysis finds that districts with a greater percentage of foodgrain area under HYV and irrigation, using greater amount of chemical fertilizer per hectare and making more intensive use of land tend to show lower relative variability of foodgrain production and yield. Thus the new technology seems to have a moderating impact on overall foodgrain production and yield variability. The introduction of biological and related innovations made it possible to replace traditional rice cultivation by a whole range of new technologies. This has important implications in terms of higher productivity per unit area as well as greater choice and flexibility. As discussed in earlier chapters (e.g. Chapter 2) the new technologies have enabled the incidence of multiple cropping to increase.

Using time series data, it was demonstrated in Chapter 9 that an increase in the incidence of multiple cropping can reduce the overall (annual) variability of production and yield and *a fortiori* reduce their coefficients of variation. This gains further support from the cross-sectional evidence provided in the preceding section of this chapter. However, we do not in this study wish to give the impression that it is the increase in the incidence of multiple cropping alone that has reduced relative variability of yields, nor that irrigation alone is the only factor involved in this. The combined package of the new technology has contributed to it and the use of many elements in the package are highly correlated. Thus in the previous section, variables such as the index of multiple cropping and the extent of irrigation might best be regarded as proxies for the introduction of a whole bundle of the new technologies.

To illustrate the correlation issue: The incidence of multiple cropping in Bangladesh is, for instance, closely associated with the availability of and expansion of irrigation, enabling greater human control to be exerted over the growing conditions of crops. Consequently both elements may add to stability. It is also conceivable that easier access to supplementary inputs such as fertilizers and pesticides, following the Green Revolution, has made it easier for growers to stabilize their production. Variations in the use of such inputs can be more finely tuned to changing environmental conditions. However, the present authors are aware that fine-tuning does not always lead to greater stability in agricultural production. Also, as pointed out earlier, experimental factors may make for a decline in relative variability of yields following the introduction of HYVs, namely, in the rejection of risky varieties after early use. Furthermore, it involves the appropriate locational-use of varieties by trial and error and in the development of and learning about appropriate cultural practices for the varieties adopted. Much more research is required to apportion the role of each of these factors in reducing yield instability.

At this stage, it is pertinent to raise some critical questions about the present analysis. To what extent has the spread of HYVs themselves been influenced by the conditions preceding the Green Revolution? To what extent has land potentially suitable for HYV cultivation influenced the adoption and diffusion of the new technology? Is it possible that lower relative variability and higher yields and the introduction of modern technology are all positively correlated? To what extent have the yields of initial years induced the adoption of HYVs? These questions are addressed below.

Detailed data for all these variables prior to the Green Revolution are not available. But given the fact that HYVs were just beginning to be introduced in the late 1960s, it would not be too unrealistic to assume that the information relating to 1969–70 is reasonably typical of conditions just prior to the introduction of the new technology. Table 10.4 presents data on overall foodgrain yield (YLD6970), percentage area irrigated in foodgrains (IRRI6970), quantity of chemical fertilizers per hectare of gross cropped area (FERT6970) for the year 1969–70 and *kharif* foodgrain area suitable (from an agricultural point of view) for HYV cultivation as a percentage of the 1969–70 to 1984–5 average *kharif* foodgrain area (POTKHYV).

A comparison of district data for the three technological proxies for 1969–70 (i.e. YLD6970, IRRI6970 and FERT6970) with those of the entire period (cf. Part A and Part B of Table 10.4) indicates that the districts show much the same rankings in the later years as in 1969–70. To see the impact of the 1969–70 values of these variables on the spread of HYVs in later years, three regression lines have been estimated with per cent area under HYV foodgrains as the dependent variable and each of the 1960–70 technological proxies as an independent variable. These are presented as Equations (10.4), (10.5) and (10.6).

$$PRHYVF = -25.6499 + 0.0388YLD6970$$

$$(R^2 = 0.5149, t = 3.990) \qquad (10.4)$$

$$\text{PRHYVF} = 12.7671 \quad + 0.8309\text{IRRI6970}$$
$$(R^2 = 0.1964, t = 1.910) \tag{10.5}$$

$$\text{PRHYVF} = 10.6394 \quad + 0.8468\text{FERT6970}$$
$$(R^2 = 0.6709, t = 5.530) \tag{10.6}$$

The estimates of Equations (10.4) and (10.6) can be considered reasonably good fits from a statistical point of view. The quality of the estimates clearly indicates considerable impact of the 1969–70 foodgrain yield and fertilizer application on the adoption of HYVs in later years. Equation (10.5) implies that the per cent of foodgrain area irrigated does not have a high explanatory power. But the coefficient is significant at 5 per cent level and has a positive sign. The high absolute values of the coefficients in Equations (10.5) and (10.6) indicate stronger impact on HYV adoption of irrigation and fertilizer application in the earlier years. Thus the conditions of the initial years are on the whole associated with the spread of the HYVs in subsequent years.

To see if POTKHYV (the area of land agriculturally potentially suitable for HYV) has any impact on PRHYVF (percentage of HYV in foodgrain area) Equation (10.7) was estimated.

$$\text{PRHYVF} = 10.3957 + 0.1690\text{POTKHYV}$$
$$(R^2 = 0.1290, t = 1.490) \tag{10.7}$$

The estimates of Equation (10.7) are poor in terms of predictive power and the coefficient is only significant at the 10 per cent level. But the positive sign of the coefficient implies that the per cent area under HYV foodgrains has a weak tendency to increase with land area potentially suitable for the cultivation of *kharif* HYV foodgrains. The data used here on the per cent area under HYV foodgrains relate to the average of the entire period from 1960–70 to 1984–5 and are considerably influenced by the lower values during the early years of the period under consideration. If one uses later period (average for the 1982–3, 1983–4 and 1984–5) data on this variable as reported in Chapter 4, a stronger relationship emerges.[5]

To examine the relationship between average foodgrain yield for the entire period (MEANYLD) and relative yield variability (CVYIELD) Equation (10.8) was estimated. The overall average yield has only a weak explanatory power and the coefficient is far from being highly significant. The negative sign, however, indicates that higher average yields tend to be associated with lower relative variability.

$$\text{CVYIELD} = 21.4144 - 0.0092\text{MEANYLD}$$

$$(R^2 = 0.1918, t = 1.890) \qquad (10.8)$$

From the above analysis, it is still possible that districts with higher initial yields had initially lower relative variability. If this were so, it might give rise to circular causation (Myrdal, 1968). But the present evidence[6] indicates that there is little systematic relationship between initial yield and initial variability of yield and production in districts and the rate of adoption of the new technology. However, there is a slight positive relationship. Possibly in districts with inherently low relative variabilities the new technology seemed less risky to farmers. This and other factors such as the extent of the land area potentially suitable for HYV cultivation and increased usage of complementary inputs, e.g. chemical fertilizers and irrigation, provided a hospitable environment for a more rapid adoption of HYVs in those districts relative to others. Furthermore, farmers in districts with higher yields at the beginning of the Green Revolution were likely to have been familiar with better crop care and modern farm management techniques ('learning by doing' factors) than farmers in districts with lower yields.

Apart from technological factors, socio-economic factors like the distribution of land might have also affected the diffusion of HYVs. Recent studies (e.g. Alauddin and Mujeri, 1986a; M. Hossain, 1980; see also Chapter 4) suggest that land concentration can be an impediment to the spread of modern technology. More equitable distribution of land in some districts than in others seems to have favoured the adoption of modern varieties of foodgrains.

10.6   CONCLUDING REMARKS

Recent studies (e.g. Hazell, 1982; Mehra, 1981) suggest that the Green Revolution has had a destabilizing impact on production and yield of foodgrains in India. As pointed out in Chapter 9, their findings might have been influenced by the assumption of arbitrary cut-off points and lack of consistency with regard to deletion of observations for 'unusual' years within a particular phase.

Examination of both national and district-level data from Bangladesh indicates that the Green Revolution may (in contrast to Hazell's findings for India) have reduced relative variability of foodgrain production and yield. Districts with higher adoption rates of HYVs and associated techniques seem to have lower relative variability. Furthermore, the probability of production and yield falling a certain per cent below the trend seem to be lower for high HYV adoption districts than those with lower adoption rates. For Bangladesh as a whole, inter-temporal analysis indicates falling relative variability of foodgrain production and yield, and this is also true for most districts in Bangladesh.

If inter-district or inter-regional data are examined, it can be found that there is a tendency for relative variability of foodgrain yield and production to fall with greater proportionate use of HYVs and with the magnitudes of the proxies for adoption of the modern technology. This may indicate that modern technology (Green Revolution') has had a moderating impact on the relative variability of yields. This study reveals that those regions with inherent lower relative variability of yields have not been more ready to adopt the new technologies, and so one can rule out this possible source of circular causation.

The evidence indicates that the Green Revolution has not in practice been a source of increased relative variability of foodgrain yield and production in Bangladesh. Bangladeshi experience indicates that on the contrary the Green Revolution may have reduced such variability. In this respect the findings for Bangladesh differ from Hazell's (1982) for India. The sets of results emerging from the employment of the alternative methodology in Chapter 9 and the use of Hazell's own method in this chapter indicate that Hazell's findings for India do not apply to Bangladesh.

As mentioned in Chapter 9, neither the quality of the statistical data used in the present analysis nor the statistical quality of the estimates themselves can justify a strong claim of establishing a view contrary to those by earlier research. Nevertheless, the results of the present study (both statistical and circumstantial) for Bangladesh provide evidence for the hypothesis that the Green Revolution has had a moderating impact on the relative foodgrain production and yield variability.

One should, however, avoid generalizing either the Indian or the

Bangladeshi experience. Rice is the predominant foodgrain in Bangladesh. In contrast, India's foodgrain production includes other important crops.[7] Rice seems to be a low-risk crop and its coefficient of variation has declined in many countries since the 1960s (Hazell, 1985). Rice is grown throughout the year and in all parts of Bangladesh. Failure of rice in one season can be cushioned, whereas other crops, e.g. wheat, barley and millets, are grown only in one season and there is little scope to cushion the effects of higher variability in one season by yields in another. This partly explains why wheat shows greater variability compared to rice in Bangladesh. Furthermore, regions in India are much more diverse in terms of climatic and geophysical characteristics than in Bangladesh. More analysis and consideration of such issues is called for, both to isolate the factors which have caused Bangladeshi experience to differ from the Indian and to identify the fundamental influences on variability which to date appear to have been mostly considered in a mechanistic fashion. Furthermore, as discussed in the next chapter, mechanistic analysis of trends in the variability of foodgrain production and yield may lull us into a false sense of security as far as the long-term risks associated with introduction of HYVs are concerned.

11 New Crop Varieties: Impact on Diversification and Stability of Yields

11.1 INTRODUCTION

In the preceding chapters, and in earlier publications, we have isolated increased multiple cropping (Alauddin and Tisdell, 1986c) and increased control over agricultural micro-environments due to greater use of HYV-associated techniques (Alauddin and Tisdell, 1988d) as significant factors contributing to reduced relative variability of crop yields with the adoption of HYVs. However, we have not given in-depth attention to the possibility that a contributor to this result may be increased crop diversification, especially greater diversity in varieties of the same crop. In this chapter we extend our earlier results by concentrating on the diversification aspect.

In so doing, we apply portfolio diversification analysis to this issue, make use of lower semi-variance and Cherbychev's inequality, and for the first time derive, from Bangladeshi data, trends in semi-variances of foodgrain yields and probabilities of disaster yield levels as well as shifts in the efficiency locus for the mean yield/semi-variance of foodgrain yields. But even if this efficiency locus shifts outward, it may not in fact be associated with greater crop diversification. We provide evidence from two Bangladeshi villages which indicates that the range of rice crop varieties grown is becoming less diverse. From a risk point of view this may not be a major problem unless existing varieties disappear. If all existing varieties remain in existence, farmers always have the option of choosing their old portfolio of cropping combinations, and if they do not choose them then, on the surface at least, they regard the new combination as superior. But if existing varieties disappear, then the available portfolio can change irreversibly and raise problems that may be overlooked in short-period diversification analysis. Let us consider these issues in the remainder of this chapter.

227

11.2   FACTORS CONTRIBUTING TO DECLINING INSTABILITY OF FOODGRAIN YIELDS

There is increasing evidence that the introduction of HYVs is associated with a fall in relative variability of foodgrain yields and has been widely observed in relation to rice. This appears to be so in Bangladesh. Three factors may help to explain this: (1) increased incidence of multiple cropping; (2) greater use of techniques such as irrigation which exert more control over crop environment; and (3) greater scope for crop diversification as new varieties are added to the new stock. We have previously provided evidence to show the significance of the two factors first mentioned (see Chapters 9 and 10; Alauddin and Tisdell, 1986c; 1988c; 1988d). We wish here to concentrate on the possible significance of the diversification aspect, to apply portfolio analysis to the subject, and in particular to consider the impact of HYV adoption on the mean yield/risk efficiency locus for foodgrains, paying particular attention to the lower semi-variance as a measure of risk or variability. To that end this section deals with conceptual issues, whereas the following section examines trends and shifts in the lower semi-variance, coefficient of lower semi-variation and shifts in the efficiency locus, mean foodgrain yields and lower semi-variance of these yields.

The Green Revolution has had a strong effect in raising average or expected yields of crops (Herdt and Capule, 1983; Alauddin and Tisdell, 1986c). This is because in some cases HYVs raise within-season yields when they replace traditional varieties, as well as increasing the scope for multiple cropping, and so on an annual basis also add to expected yields. If a single crop of an HYV has a lower yield than a single crop of a traditional variety, and if the former permits multiple cropping but the latter does not, annual expected yield may be much higher with HYV introduction. To the extent also that yields between seasons are not perfectly correlated, this will tend to reduce risks by, for example, lowering the coefficient of variation of annual yields, even though as we have discussed elsewhere (Alauddin and Tisdell, 1987; see also Boyce, 1987a) production sustainability problems may emerge in the long term. Multiple cropping is likely to reduce the probability of annual farm income falling below a disaster level, if we leave the secular problem to one side (Anderson *et al.*, 1977, p.211).

While HYVs may have a higher expected yield and greater risk or yield variability than traditional varieties, in some cases HYVs may

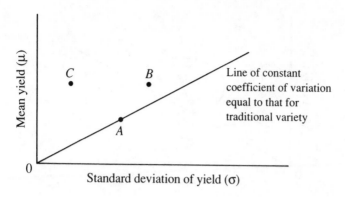

Figure 11.1 HYVs often appear to have a lower coefficient of variation of yield than traditional varieties and sometimes have a lower standard deviation and a higher mean yield

involve higher yields and fewer risks to individual farmers than traditional varieties. In the latter case, they dominate traditional varieties and can be expected to replace them in due course. As explained in the next section, techniques associated with HYVs, by ensuring greater environmental control, may help to establish this dominance.

As is well known, there is no simple shorthand way of measuring uncertainty and instability. But for the purpose of this exercise let us take the variance or the standard deviation as an indicator. In Figure 11.1, the mean income for a farm and its standard deviation are shown. The combination at *A* may indicate the situation of the farm using a traditional variety. The advent of an HYV may, if it replaces the traditional variety, make *B* or even *C* possible. Aggregate evidence from Bangladesh indicates a rise in absolute variability but a reduction in relative variability following the Green Revolution, as well as a rise in mean yields (Alauddin and Tisdell, 1988c; 1988d). So the overall situation is depicted by a point to the right of *A* located above line *OA*. The slope of *OA* is determined by the mean yield of the traditional variety divided by its standard deviation. In the case illustrated, the yield characteristics of HYV at *B* lie above the line *OA* and reduce the coefficient of variation, even though the HYV has a much greater variance than the traditional variety.

Where the variability of yields from HYVs is greater than that for traditional varieties, diversification of varieties grown can be used as a strategy to reduce the risk of growing some HYVs, provided yields

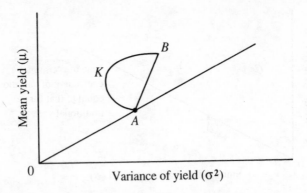

Figure 11.2   As the efficiency loci indicate, the availability of HYVs may permit diversification of variety of crops grown and reduce relative variability of yields on farms, at least in the short run

of varieties are not perfectly correlated. If the yields of the varieties are perfectly correlated, and if A and B are the unmixed crop variety possibilities, the efficiency locus (Markowitz, 1959) is indicated by the line AB in Figure 11.2 (Anderson *et al.*, 1977, p.193). Lack of perfect correlation between returns from the different crop varieties results in the efficiency locus or curve joining A and B to bulge to the left so that a curve like AKB in Figure 11.2 may result (for details on the nature of this curve see Anderson *et al.*, 1977, p.193). Thus when returns are not perfectly correlated, variety diversification can reduce risk and lower the coefficient of variation.

Cherbychev's inequality and variations on it can be used to explore this matter further (Anderson *et al.*, 1977, p.211). According to this inequality the probability that a random variable, x, deviates from its expected value, μ, by more than an amount k is equal to or less than its variance divided by k, that is

$$Pr\left(|x - \mu|\right) \geq k) \leq \sigma^2/k \tag{11.1}$$

According to the inequality, if an income of x_1 or less than x_1 is to be avoided (would be a disaster), the probability of not avoiding this is given by the following inequality

$$Pr\left(x \leq x_1\right) \leq \sigma^2/\left(\mu - x_1\right)^2 \tag{11.2}$$

A farmer may require the probability to be less than a particular value, say k. This will be satisfied if

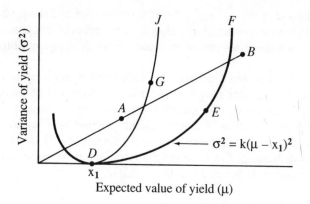

Figure 11.3   Curves based on Cherbychev's inequality (and a modification to it based on lower semi-variance) indicating combinations which satisfy and which do not satisfy 'safety-first' requirements

$$\sigma^2 / (\mu - x_1)^2 \leq k \tag{11.3}$$

Thus if σ^2 is assumed to be the dependent variable, all combinations of (μ, σ^2) on or below the parabola

$$\sigma^2 = k(\mu - x_1)^2 \tag{11.4}$$

satisfy this constraint, except for all combinations involving a value of $\mu < x_1$, x_1 being assumed to be less than μ. Thus for the case illustrated in Figure 11.3 combinations to the left of the *DEF*, the positive branch of relevant parabola may not satisfy the safety-first constraint, whereas those to the right of this branch or on it do satisfy the constraint.

Several points may be noted:

1. Where returns from different varieties are not correlated, and on the basis of the argument illustrated in Figure 11.2, it may be possible to meet the safety-first constraint by diversification of varieties grown.
2. If the variance and mean level of income increase in the same proportion, the likelihood of the safety-first constraint being satisfied rises. For example, if in Figure 11.3, *A* is the combination of (μ, σ^2) for the traditional variety and *B* corresponds to that for an HYV, the HYV meets the constraint but the traditional variety does not. As one moves out further along the line *OA*, the

likelihood of the constraint being satisfied rises and the probability declines of incomes falling below x_1. This can even happen, up to a point, when μ rises and σ^2 increases more than proportionately.

It should be noted Cherbychev's inequality is not very powerful. Because of this, there *may* be combinations to the left of curve *DEF* in Figure 11.3 which also satisfy the probability constraint. In that respect a modification of this inequality so that it is based on the lower semi-variance is more powerful (Tisdell, 1962). If the probability distribution of income is symmetric about its mean the lower semi-variance is $0.5\sigma^2$ and it follows that

$$Pr\ (x < x_1) \leqslant 0.5\sigma^2/(\mu-x_1)^2 \tag{11.5}$$

The relevant safety-first parabola now is

$$\sigma^2 = 2k(\mu-x_1)^2 \tag{11.6}$$

and the relevant branch might be as indicated by curve *DGJ* in Figure 11.3. Clearly similar consequences follow to those mentioned earlier.

11.3   LOWER SEMI-VARIANCE ESTIMATES AND SHIFTS IN MEAN/RISK EFFICIENCY LOCUS FOR FOODGRAIN YIELDS IN BANGLADESH

The lower semi-variance and the modified Cherbychev inequality as set out in Expression (11.5) provides a more relevant measure of the risk from yield variability faced by the farmers than does the variance and Cherbychev's inequality. We used the variance in our earlier analysis (Alauddin and Tisdell, 1988c) but we wish in this section to consider changes in the lower semi-variance using Bangladeshi data for foodgrain yields, and also to consider trends in the probability of a disaster yield in Bangladesh, using the modified Cherbychev inequality. This will lead on to estimates of shifts in the mean/risk foodgrain yield efficiency frontier for Bangladesh, a matter which does not seem to have been previously considered.

Let us now consider empirical estimates using Bangladeshi foodgrain yield data. First of all, annual indices of overall foodgrain yield were derived using the average of the triennium ending 1977–8 as the

base. In measuring the degree of variability, the present chapter
follows the approach used in one of our earlier studies (Alauddin and
Tisdell, 1988c). As the primary concern of this chapter lies in the
deviations below the mean, we choose a period of longer duration
(than the five years employed by Alauddin and Tisdell, 1988c). This
is because a five-yearly period is likely to reduce significantly the
number of observations for negative deviations and is unlikely to
provide a robust basis of analysis. Using this approach and applying it
to Bangladeshi time series data for the period 1947–8 to 1984–5, we
specify the changing behaviour over time of yield variability below
the mean. Moving values for 9 and 11 years were tried and the results
were similar. The results presented in this chapter are based on the
11-year period. The relevant data are set out in Table 11.1.

A few observations seem pertinent. First, there is a steady increase
in foodgrain yields. Secondly, there seems to be little *overall* time
trend in lower semi-variance, (lower) coefficient of semi-variation
and probability of yield falling 50 per cent or more below the initial
11-year average yield. However, these measures of variability seem
to increase initially and then decline before showing a rising tendency
once again at a slower rate than initially. Indeed at about the
mid-1960s there appears to be a fundamental change in relative
variability of yields. A large and significant downward shift occurs in
the trend of relative variability, and it increases at a slower rate than
prior to the mid-1960s.

These aspects come into sharper focus when one plots the relevant
observations against time. Figure 11.4 plots (lower) semi-variance
(SEMVAR). One can visually identify two distinct phases (the first
phase 1952–63, that is, 1952–3 to 1963–4, and the second phase
1964–79, that is, 1964–5 to 1979–80). Up to the early 1960s (lower)
semi-variance of yield shows a strong tendency to increase. In order
to make quantitative comparison of change in its behaviour, three
regression lines with time (T) as the explanatory variable were
estimated. These are presented as Equations (11.7), (11.8) and (11.9)
for the corresponding periods. Figure 11.4 illustrates Equations (11.7)
and (11.8) for the first and second phases respectively. Equation (11.7)
clearly indicates a strong time trend in (lower) semi-variance in terms
of both explanatory power and statistical significance. Equation
(11.8) shows a similar trend in the later period. However, the relative
rate of change is much lower in the second phase compared to the
one in the first, as indicated by their respective slopes. More

234

Table 11.1 Mean values, lower semi-variance, coefficient of semi-variation of yield of all foodgrains, and probability of yield falling below a 'disaster level': Bangladesh, 1947–8 to 1984–5

Year	Index	MEANYLD	LAVERAG	SEMVAR	SEMCOV	PROB50
1947	70.74					
1948	78.79					
1949	75.44					
1950	73.29					
1951	69.23					
1952	70.50	73.231	69.386	5.441	3.362	0.00406
1953	74.80	73.195	69.305	5.203	3.291	0.00389
1954	71.06	73.304	69.970	6.807	3.729	0.00506
1955	65.40	74.332	69.970	6.807	3.729	0.00478
1956	81.34	75.853	70.899	10.874	4.651	0.00706
1957	74.96	76.925	71.178	12.396	4.946	0.00763
1958	70.34	79.032	71.314	15.357	5.495	0.00854
1959	79.98	80.452	72.349	29.751	7.539	0.01550
1960	86.75	82.096	75.507	42.678	8.652	0.02060
1961	90.02	83.783	77.527	22.724	6.149	0.01020
1962	81.02	84.761	78.053	29.112	6.913	0.01260
1963	93.68	86.492	78.826	34.836	7.488	0.01400
1964	90.42	88.512	82.928	9.345	3.686	0.00347
1965	89.14	89.374	85.218	12.327	4.120	0.00443
1966	83.96	89.222	84.800	11.317	3.967	0.00409
1967	92.09	88.599	86.606	2.946	2.060	0.00109

Year						
1968	94.00	89.939	86.163	8.686	3.420	0.00305
1969	92.57	89.787	86.163	8.686	3.420	0.00307
1970	89.46	90.540	86.163	8.686	3.420	0.00299
1971	85.08	91.098	86.418	7.882	3.287	0.00266
1972	83.17	93.104	89.063	16.185	4.517	0.00507
1973	95.75	94.233	89.681	19.108	4.891	0.00576
1974	92.01	94.958	88.467	17.482	4.727	0.00514
1975	98.70	96.493	90.125	27.217	5.789	0.00759
1976	95.28	97.977	90.267	33.891	6.450	0.00900
1977	106.03	100.230	92.982	35.737	6.429	0.00883
1978	104.52	102.953	96.742	14.167	3.891	0.00322
1979	101.97	104.944	98.495	25.196	5.096	0.00540
1980	109.46					
1981	105.79					
1982	109.85					
1983	113.12					
1984	117.66					

Notes: 1947 means 1947–8 (July 1947 to June 1948), etc. Index numbers are constructed with the average of the triennium ending 1977–8 = 100. Mean values are 11-yearly moving averages of the index numbers. SEMVAR is lower semi-variance. LAVERAG is average of indices below their mean. SEMCOV = [√(SEMVAR/LAVERAG)] × 100. PROB50 is probability of yield falling 50 per cent or more below the first of the 11-yearly moving mean yield (i.e. 73.231).

Source: Based on Table 9.1

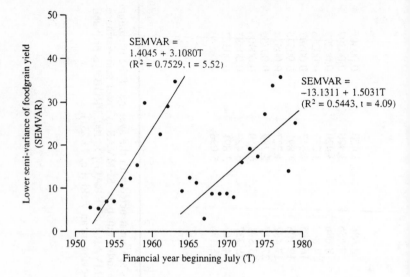

Figure 11.4   Lower semi-variance (SEMVAR) of foodgrain yields in Bangladesh, based on 11-year moving averages, 1947–8 to 1984–5. Notice the downward shift in the apparent trend function about the mid-1960s

importantly, there is a change in trend apparent with the relative variability indicated by Equation (11.8) being much lower than that for (11.9) as a function of time.

Phase 1:   SEMVAR = $1.4045 + 3.1080T$,
$$R^2 = 0.7529, \ t\text{-value} = 5.52 \qquad (11.7)$$

Phase 2:   SEMVAR = $-13.1311 + 1.5031T$,
$$R^2 = 0.5443, \ t\text{-value} = 4.09 \qquad (11.8)$$

Entire :   SEMVAR = $11.7585 + 0.4011T$,
$$R^2 = 0.0895, \ t\text{-value} = 1.60 \qquad (11.9)$$

Figure 11.5 plots coefficients of (lower) semi-variation (SEMCOV) against time. One can identify two similar phases in its behaviour to those for (lower) semi-variance. To facilitate quantitative comparisons, Equations (11.10), (11.11) and (11.12) were estimated by least squares linear regression. Equations (11.10) and (11.11) have been illustrated in Figure 11.5. Equation (11.10) corresponds to the first phase and shows that the coefficient of (lower) semi-variation had a strong tendency to increase during Phase 1. The strong explanatory power and high *t*-value lend clear support to this claim. However, there is an important change in its behaviour when one considers

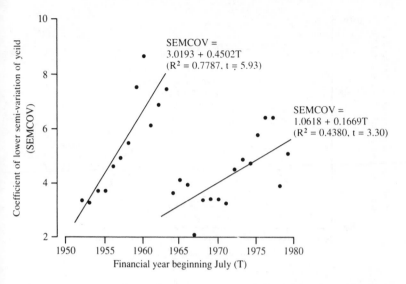

Figure 11.5   Coefficient of lower semi-variation (SEMCOV) of foodgrain yields in Bangladesh, based on 11-year moving averages 1947–8 to 1984–5. Notice the strong downward shift in the apparent trend function about the mid-1960s and the much flatter slope of the trend line of the coefficient of variation in the second phase

Equation (11.11) which relates to the second phase and Equation (11.12) which relates to the entire period. A comparison of Equations (11.10) and (11.11) clearly shows a much slower rate of increase in the coefficient of (lower) semi-variation in the second phase. Once again a strong downward shift in relative variability is suggested by the comparison of Equations (11.10) and (11.11) and their lines shown in Figure 11.5.

Phase 1:   SEMCOV = 3.0193 + 0.4502T,
$$R^2 = 0.7787, \text{ } t\text{-value} = 5.93 \qquad (11.10)$$
Phase 2:   SEMCOV = 1.0681 + 0.1669T,
$$R^2 = 0.4380, \text{ } t\text{-value} = 3.30 \qquad (11.11)$$
Entire :   SEMCOV = 4.7619 + 0.0047T,
$$R^2 = 0.0006, \text{ } t\text{-value} = 0.12 \qquad (11.12)$$

Figure 11.6 depicts the behaviour of probability of yield falling 50 or more below the average of the initial 11-year period (PROB50). Overall no time trend can be established. But two distinct phases can be identified. It increases during the period up to the early 1960s

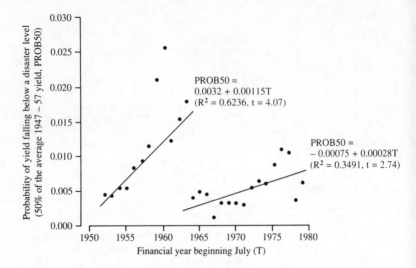

Figure 11.6   Probability of foodgrain yield falling below a disaster level (PROB50) (arbitrarily defined as 50 per cent of the average yield 1947–57) for Bangladesh, based on 11-year moving averages, 1947–8 to 1984–5

(Phase 1) and falls to lower values on average in the subsequent period (Phase 2). The behaviour of PROB50 can be placed into a pattern by considering the estimated regression equations in Figure 11.3 and Equations (11.13), (11.14) and (11.15) which relate respectively to the first and second phases and the entire period. A comparison between Equations (11.13) and (11.14) and the corresponding lines shown in Figure 11.7 suggests that the trend in probability of a disaster level of yield had undergone a significant downward shift around the mid-1960s.

Phase 1: $PROB50 = 0.00320 + 0.00115T,$
$R^2 = 0.6236, t\text{-value} = 4.07$ (11.13)

Phase 2: $PROB50 = 0.00075 + 0.00028T,$
$R^2 = 0.3491, t\text{-value} = 2.74$ (11.14)

Entire : $PROB50 = 0.00824 - 0.00011T,$
$R^2 = 0.0420, t\text{-value} = 1.07$ (11.15)

Figure 11.7 plots (lower) semi-variance against mean yield (MEANYLD). It follows a similar pattern to those in Figures 11.4–11.6.

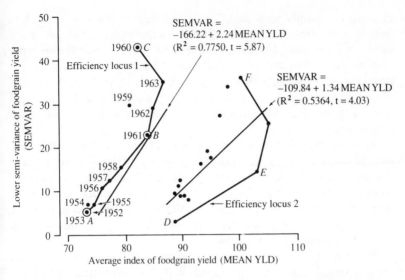

Figure 11.7 Scatter of lower semi-variance (SEMVAR) of foodgrain yield and mean value of foodgrain yield (MEANYLD) for Bangladesh, based on 11-year moving averages, 1947–8 to 1984–5. Note the apparent shift in scatter after 1963 (centre point) and the strong shift rightward in the hypothetical efficiency locus

Note that the scatter falls into two distinct clusters or sets, with observations centred on 1963 or earlier being well to the left of those for the later period. If a linear least squares approximation is made to the two clusters, the following equations are obtained:

Cluster 1: SEMVAR $= -166.2208 + 2.2395$ MEANYLD,
$R^2 = 0.7750$, t-value $= 5.87$ (11.16)
Cluster 2: SEMVAR $= -109.8443 + 1.3425$ MEANYLD,
$R^2 = 0.5364$, t-value $= 4.03$ (11.17)

These indicate a significant shift downward in the (lower) semi-variance in relation to mean foodgrain yield about the mid-1960s, and that the (lower) semi-variance is increasing at a slower rate in relation to the mean level of foodgrain yield in the second phase than in the first. In addition, *if it is assumed* that Cluster 1 is drawn from a distinct population, the efficiency locus in Phase 1 is as indicated by the heavy line segments passing through the points such as *ABC*, whereas in Phase 2 it is indicated by heavy line segments passing through the points such as *DEF*. This would imply that there has been a considerable outward shift (i.e. to the right) in the efficiency locus. All the available evidence strongly points towards this conclusion. Note, however, the efficiency loci as shown are rough approximations, and more than two loci are likely to have applied in the period under consideration. Also such loci, unlike Locus 1, should be non-reentrant. But the order of change is such as to override such considerations.

From the available data, it seems that relative variability of Bangladeshi foodgrain production shifted downwards significantly around the mid-1960s. While the relative variability below the mean is still continuing to rise, it appears to be doing so at a much slower rate than prior to the early 1960s. Similar shifts have occurred in relation to the probability of a 'disaster' level of yield, meaning the shift was downward around the mid-1960s.

Why does the apparent break occur in trends in variability? The second period, 1964–79, corresponds with the commencement of the introduction of the new technology *to control* agricultural production. Increased irrigation and fertilizer use, followed later by the introduction of HYVs, occurred during the second phase, as distinguished here. Note, however, that in both phases there is a tendency for relative variability of yield below the mean to increase with time. The reason for this trend is unclear at present but is worthy of future investigation.

Nevertheless, it should be emphasized that this whole analysis depends on mathematically *expected* values (averaging procedures) even when it takes account of higher moments. While there may be a reduction in the *relative frequency* or *probability* of disaster, disaster when 'disaster' comes may be more *catastrophic*. This is not captured by averaging procedures. Thus, while use of HYVs may reduce the probability of a disaster level of income (or less) occurring, a catastrophically lower income may occur when disaster strikes. More research into the probability of this is needed.

11.4 EVIDENCE FROM BANGLADESH ABOUT THE EXTENT OF CROP DIVERSIFICATION FOLLOWING THE 'GREEN REVOLUTION'

As observed in the last section, the mean/lower semi-variance of foodgrain yield in Bangladesh appears to have shifted to the right with the introduction of HYVs. While at least in the short run the introduction of new varieties makes greater crop diversification possible, there is no guarantee that increased diversification will occur. Indeed, if the expected return/risk combination of the new varieties is sufficiently favourable, less diversification both of crops and varieties of crops can occur. In fact, the limited field evidence which we have points towards less diversification following the Green Revolution.

Before considering that evidence, note that in Bangladesh the Green Revolution is confined to cereals (rice and wheat) and partly to jute and sugar cane. Even though some success in research in other crops like potato, summer pulses and oilseeds is believed to have been achieved (Alauddin and Tisdell, 1986a; Gill, 1983; Pray and Anderson, 1985) it is yet to take off in any real sense. Recent evidence (e.g. Alauddin and Tisdell, 1988g) indicates that the area under rice and wheat has expanded at the expense of non-cereals. This substitution of cereals for non-cereals has resulted from, among other things, the improvement in the technology specific to the former, relative to the latter. The output of non-cereals as a whole has declined as a result of decline in yield and hectarage. Thus there is a comparative 'crowding out' of non-cereal crops by cereal production following the Green Revolution. On the other hand, cultivated land that was once left fallow for a significant part of each year (for example, during the dry season) is now used for crops such as wheat and dry-season rice varieties. Thus the Green Revolution has induced greater *monocultural* multiple cropping.

Let us consider some farm-level evidence based on the survey reported in Chapter 6. An analysis of the farm-level data indicates that the two villages exhibit very different cropping patterns. Ekdala has a more traditional cropping pattern. Apart from growing cereals (rice and wheat), a number of other crops including sugar cane, pulses, jute and oilseeds, and fruit crops such as banana and watermelon, are also grown. Cereals occupy 65 per cent of gross cropped area. Overall cropping intensity is quite high (over 178 per cent) and is considerably in excess of the national figure of 150 per cent in recent years (Alauddin and Tisdell, 1987).

South Rampur on the other hand has a much more specialized cropping pattern, in that rice is the only crop grown. Almost every plot of land is double cropped with rice. Intensity of cropping is significantly higher (nearly 200 per cent) than that in Ekdala. Furthermore, it is observed from the Ekdala data that farmers with access to irrigation allocate a significantly higher percentage of gross cropped area to rice and wheat than those without access to irrigation. The latter category of farmers allocate a significantly higher percentage of gross cropped area to non-cereals like sugar cane, pulses, oilseeds, watermelon. Irrigated farms cultivate land much more intensively than non-irrigated farms where cropping intensity is much lower (about 116 per cent) than the overall intensity of cropping for Bangladesh.

Thus the availability of irrigation has considerable impact on cropping pattern and intensity of cropping. A plot of land under irrigation is normally double cropped with rice (*boro* HYV rice followed by local or HYV *aman* rice). This, however, involves (rice) monoculture multiple cropping. This seems to be consistent with the changes in cropping pattern for Bangladesh as a whole. This trend is likely to be causes for concern.

Firstly, monocultural multiple cropping of cereals may prove to be ecologically unsustainable or less sustainable than in the past, and there are already signs that it is becoming more costly and difficult to maintain productivity using such cultural practices (Hamid *et al.*, 1978, p.40; Alauddin and Tisdell, 1987).

Secondly, as discussed in Chapter 5, mere production of increased quantities of rice at the expense of other non-cereal food crops, e.g. pulses, vegetables, is unlikely to solve food and nutritional problems.

It would be useful to have data for Bangladesh indicating whether the number of strains or varieties of crops such as rice grown on farms is increasing or decreasing, and have some information about trends in the total number of varieties available. It is possible that with the introduction of HYVs the number of available varieties at first increases and then declines. At the same time, the *fundamental* gene bank may be declining while the number of available strains or varieties is at first increasing. So a difficult measurement problem in relation to genetic diversity exists.

Only limited evidence is available from our survey areas of Ekdala and South Rampur. The farm-level data from Ekdala indicate that farmers primarily rely on two or three varieties of HYV rice. During the 1985–6 crop year, BR11 and China varieties constituted 82 per cent of the gross area planted to HYV rice. In South Rampur, farmers were found to allocate over 70 per cent of HYV rice area to four rice varieties: BR3 (21 per cent), BR11 (26 per cent). *Paijam* (13 per cent) and Taipei (10 per cent). Furthermore, the South Rampur Survey indicated that the number of rice varieties in use has fallen from about 12–15 in the pre-Green Revolution period to 7–10 in the post-Green Revolution phase. This also seems to be supported by the Ekdala evidence.

11.5   PROBLEMS OF REDUCED GENETIC DIVERSITY OF CROPS: A SECULAR DECLINE IN AVAILABLE VARIETIES AND STRAINS

Our limited evidence from Bangladesh indicates that the introduction of HYVs has been accompanied by less crop diversification on farms; even though the relative variability of foodgrain yield has fallen, its mean/risk efficiency locus has become more favourable. The latter trends thus cannot be attributed to any significant extent to greater diversification of crops.

While a favourable reduction in yield risks appears to have accompanied the Green Revolution, these trends may disguise long-term rising risk. In the long term, the introduction of HYVs may paradoxically reduce the scope for diversification. This problem would not arise if varieties in existence prior to the introduction of HYVs continue to be available, along with the development of other improved varieties. But improved varieties quite frequently drive existing varieties out of usage and existence, so that in the long term less choice of varieties may be available than prior to the introduction of HYVs (cf. Plucknett *et al.*, 1986, esp. pp.8–12). Furthermore, the varieties which disappear may be those which provide the most valuable genetic building blocks for development of new varieties. Thus crop productivity may become dependent on a limited number of varieties and the risk of production being unsustainable may rise considerably (cf. Plucknett *et al.*, 1986).

New varieties of crops appear to have a limited life, on average. The World Conservation Strategy (WCS) document (IUCN, 1980) suggests that wheat and other cereal varieties in Europe and North America have a lifetime of only 5–15 years.

This is because pests and diseases evolve new strains and overcome resistance; climates alter, soils vary; consumer demand changes. Farmers and other crop-producers, therefore, cannot do without the reservoir of still-evolving possibilities available in the range of varieties of crops, domesticated animals, and their wild relations. The continued existence of wild and primitive varieties of the world's crop plants provides humanity's chief insurance against their destruction by the equivalents for those crops of chestnut blight and Dutch elm disease (IUCN, 1980, sec. 3.3).

The WCS document goes on to point out how the bulk of Canadian wheat production now depends on four varieties, as does most of US potato production, and provides other examples of growing agricultural dependence on a narrower range of varieties. Furthermore, there appears to have been a rapid disappearance of primitive cultivars, for instance, the percentage of primitive cultivars in the Greek wheat crop fell from over 80 per cent in the 1930s to under 10 per cent by the 1970s, and the absolute number of these declined quite considerably (IUCN, 1980, sec. 3.4; Allen, 1980, p.41). Such declines are claimed to be 'typical of most crops in most countries' (IUCN, 1980, sec. 3.4). Thus while in the shorter term new varieties may become available and increase the scope for reducing instability of production and income, for instance via diversification, in the long term the opposite tendency may be present.

This raises a number of questions. For instance, if varieties disappear, is some market failure present and will the disappearance lead to a socially sub-optimal outcome? Is government intervention required to correct the situation? If so, what guidelines should be adopted in determining which crop varieties to preserve? What social mechanisms should be adopted to bring about the range of conservation of varieties required?

These complex questions cannot be addressed in detail here. However, there is reason to believe that market failure and transaction costs can lead to the socially sub-optimal disappearance of species (see Tisdell and Alauddin, 1988b). Furthermore, a variety of criteria have been suggested for determining which array of species and varieties to conserve. These range from approaches using cost-benefit analysis (CBA) to those advocating a safe minimum standard (SMS) (Bishop, 1978; Chisholm, 1988; Ciriacy-Wantrup, 1968; Quiggin, 1982; Randall, 1986; Smith and Krutilla, 1979).

11.6 CONCLUDING COMMENTS

The available evidence from Bangladesh points to a major decline in the relative variability of foodgrain yields following the introduction of new crop technology associated with the Green Revolution. An analysis of mean yields and (lower) semi-variance data for Bangladesh indicates a strong shift downwards in the trend line of relative instability of crop yields from about the mid-1960s, and a major break in trends with rates of increase in yield instability being lower after the mid-1960s. Factors which may be responsible for this phenomenon were discussed, and particular attention was given to the diversification of crops and varieties of particular crops grown as a possible contributor. Our data from Bangladesh are inadequate at present to determine whether there is more (or less) crop diversification following the Green Revolution. While initially the advent of the new varieties would seem to expand the available choice of varieties and possibilities for crop diversification for farmers, this may not be so in the long term. Some traditional varieties may be dominated *initially* by some favourable characteristics of new varieties. Consequently these traditional varieties may disappear, thereby reducing available genetic diversity and raising risks in the long run (cf. Plucknett *et al.*, 1986, Ch. 1). This raises a possible dilemma and a basic issue for policy – namely, what is the responsibility of governments to preserve genetic diversity and how should governments decide which varieties to preserve? Economists are divided in their advice to governments about the appropriate decision-making model to apply to such choice problems (cf. Randall, 1986; Tisdell, 1990a; 1990b).

12 Trends in and Projections of Bangladeshi Food Production and Dependence on Imported Food

12.1 INTRODUCTION

In earlier chapters, using official Bangladeshi statistics, we indicated that Bangladeshi foodgrain production grew at a substantially faster rate following the Green Revolution than prior to it. However, Boyce (1985) argues that there are a number of shortcomings in the official agricultural production statistics, and he derives an alternative data series for agricultural crops. In the light of this, we derive an alternative data series for *foodgrain* production based on the Boyce's (1985) assumptions, and use it and the official data to estimate growth rates for various sub-periods since the early 1950s. However, we raise serious doubts about the qualitative superiority of the revised data series over the official data, while acknowledging Boyce's contribution.

We then consider trends in the availability of foodgrain from domestic production as well as its import intensity over the years. This highlights the fact that Bangladesh has become more dependent on food imports following the Green Revolution and from that point of view its food security has been reduced. This, however, is not to argue that the connection is necessarily a causal one, even though that possibility cannot be dismissed entirely.

12.2 REVIEW OF THE EVIDENCE ON PAST GROWTH AND THE VIEW OF BOYCE

Any analysis of agricultural production in Bangladesh encounters formidable problems of data quality. The only comprehensive and

246

major source of data is the Bangladesh Bureau of Statistics. Questions have been raised about the reliability of the data provided by the official source. Pray (1980) made an extensive survey of existing sources of agricultural data and identified sources of errors and weaknesses of these data systems. Pray argued that the employment of the official data can give a misleading picture of the level and trend in output and can result in faulty policy advice. Clay (1986, p.178) points out that the secondary data of agricultural production or any other policy-sensitive area in Bangladesh must be treated with the greatest circumspection'. Boyce (1985, p.A31), 'by means of comparisons with alternative published and unpublished data sources' has identified 'a number of systematic errors in the official crop acreage and yield series' and prepared a revised series of agricultural output and yield. The use of his revised series leads to a reversal in the trends of growth rates in agricultural output in that 'whereas agricultural output growth rates estimated from official data indicate a decline from 2.15 per cent per annum in the period 1949–64 to 1.52 in the period 1965–80, the revised data series reveals an opposite movement: output growth rose from 1.27 per cent in the earlier period to 2.18 per cent in the latter' (Boyce, 1985, p.A31).[1]

As it is important for the present discussion to obtain as accurate a picture of trends in Bangladeshi food production as is possible, Boyce's procedure for deriving a revised series for agricultural crop area, yield and output is considered. The revised data series is presented in terms of indices (1949–50 = 100) derived from estimated crop output measured in value terms (Boyce, 1985, p.A40). Boyce's study encompasses all major agricultural crops including non-food crops such as jute. Since rice and wheat are the main sources of food in Bangladesh, the present study will only concentrate on these foodgrains. First of all it briefly outlines the way in which Boyce revises the official foodgrain data. It is worth mentioning here that Boyce does the revision only in the case of *aus* and *aman* rice crops, as he did not find adequate alternative evidence upon which to base a revision of the official *boro* rice and wheat statistics (Boyce, 1987a, p.333).

The salient features of the Boyce revision are summed up as follows (for details see Boyce, 1985, p.A40):

1. *Aman* rice: *Aman* area (1964–5 to 1966–7) and yield (1962–3 to 1966–7) statistics are revised to allow for the objective estimates by multiplying the official data by annual scalars derived as the ratio of objective to official estimates (Boyce, 1985, p.A33).

Revised yield estimates for the 1959–60 to 1961–2 period are derived, using the adjustment scalar defined as the average ratio of objective to subjective estimates for the 1962–3 to 1966–7 period. Output estimates for the 1967–8 to 1969–70 period have been derived from Alamgir and Berlage (1974, pp.20, 183), taking the official area figures as accurate and absorbing the adjustment on the yield side.

2. *Aus* rice: *Aus* area (for 1964–5 to 1967–8) and yield (for 1962–3 to 1967–8) are revised, using adjustment scalars similar to those in case of the *aman* crop. Revised yield estimates are derived for the 1959–60 to 1961–2 period, using the adjustment scalar defined as the average ratio of objective to subjective estimates for the 1962–3 to 1966–7 period. *Aus* area figures for the 1949–50 to 1963–4 and the 1970–1 to 1980–1 periods are revised using cumulative adjustment factors akin to a compound interest rate which rises from 2.2 per cent in 1949–50 to 8.6 per cent in 1963–4. For the latter period it rises from 1.1 per cent in 1970–1 to 12.4 per cent in 1980–1.

Employing Boyce's approach, revised statistics of the *aman* and *aus* area, yield and production of rice and wheat were derived. Thus the revised foodgrain statistics (hereinafter *revised series*) is inferred by the present authors using Boyce's hypothesis as applied to the rice component of agricultural production. The *revised series* is set out in Table 12.1 alongside the corresponding official data. A comparison of the two data series indicates that the official data slightly but consistently 'underestimate' foodgrain area and production, but there is hardly any difference between the two yield series in the 1950s. For the major part of the 1960s, the official sources report foodgrain areas which are slightly lower than those obtained by the Boyce revision. The official production figures are significantly higher, resulting from 'reported overestimation' of yields in the official series during the 1960s. During the 1970s, however, the official area and production figures are generally but not consistently lower than those resulting from the revision. Yields reported in the official series are marginally and consistently higher than those in the revised series. Changes in the growth rates reported by Boyce are due primarily to adjustment of the official production estimates to correct for reportedly 'overestimated' yields in the late 1950s and the 1960s.

Boyce strongly argues that the official data to underestimate yield in recent years. Yet the *revised series* obtained using his assumptions indicate that the revised yield estimates differ little from the official series. If anything, the revised estimates are *slightly lower*. The Boyce hypothesis, therefore, does not appear to be substantiated by the *revised series* derived employing his assumptions.

The validity of application of cumulative adjustment factors to the official statistics for the period 1949–50 to 1963–4 and for 1970–1 to 1980–1 is crucial for the revision of the *aus* area under cultivation. Assuming that *aus* area has been underestimated in the official statistics in the above periods, there is no guarantee that it would in fact assume the values resulting from the adjustment factors employed by Boyce. If the revised area figures were to differ from their 'true values', production figures would likewise differ. Given the generally poor production performance of Bangladeshi agriculture, the growth rates estimated from the Boyce data might well give a spurious indication of the level and trend in foodgrain production.

On closer examination two inconsistencies can be identified in the revised series. First, upon adjustment one finds that the 1969–70 foodgrain output is lower than that of 1970–1. In reality, however, this does not seem plausible because 1969–70 was the last normal year before the War of Liberation, which occurred in the last quarter of 1970–1 and is likely to have affected harvest of *boro* rice and maybe part of *aus* rice. Losses of rice crops during 1969–70 were estimated at 223 thousand tonnes as against 1981 thousand tonnes in 1970–1 (BBS, 1978, p.156). It might be noted that the Boyce index for agricultural output for 1969–70 is marginally higher than that of 1970–1 (Boyce, 1985, p.A40). This is probably because agricultural output is measured in value terms and is an aggregate for all crops rather than foodgrains alone. Relative prices can conceal underestimation in quantitative terms.

Secondly, Boyce (1985, p.A32) criticizes the official data for indicating that 1978–9 *aman* output rose marginally compared to that of 1977–8, even though 1977–8 was generally acknowledged to have been a 'bumper year' while 1978–9 recorded a 'poor harvest'. The official sources (e.g. BBS, 1985b, p.303) put the estimated losses of rice output due to flood and drought at 112 and 102 thousand tonnes for 1977–8 and 1978–9 respectively. Consider the revised production figures for 1977–8 and 1978–9. It can be seen that the production level for the 'poor' year is slightly higher than that of the 'bumper' year. Once the wheat outputs for the two years are deducted from the corresponding total foodgrain production, the rice output in 1977–8 is seen to be slightly higher (about 37 thousand tonnes) than that in

Table 12.1 Official[a] and revised[b] foodgrain statistics: Bangladesh, 1947–8 to 1982–3

Year	Area ('000 ha)			Production ('000 tonnes)			Yield (kg/ha)		
	Official	Revised	% Diff.	Official	Revised	% Diff.	Official	Revised	% Diff.
1947	7725.89	NA	–	6864.21	NA	–	888	NA	–
1948	7899.06	NA	–	7815.33	NA	–	989	NA	–
1949	7941.75	7980.00	-0.479	7518.34	7543.57	-0.334	947	945	0.179
1950	8134.22	8188.66	-0.665	7481.15	7526.12	-0.598	920	919	0.099
1951	8253.96	8317.07	-0.759	7170.75	7217.26	-0.664	869	868	0.142
1952	8448.01	8512.70	-0.760	7477.19	7525.25	-0.639	885	884	0.113
1953	8946.91	9028.48	-0.903	8401.08	8469.78	-0.811	939	938	0.094
1954	8675.97	8761.95	-0.981	7737.60	7808.05	-0.902	892	891	0.098
1955	7923.83	8017.92	-1.173	6509.20	6582.66	-1.116	821	821	0.000
1956	8169.95	8275.75	-1.278	8339.61	8436.07	-1.143	1021	1019	0.160
1957	8232.07	8346.13	-1.367	7742.98	7845.47	-1.306	941	940	0.105
1958	7989.26	8114.11	-1.539	7057.97	7144.12	-1.206	883	880	0.289
1959	8615.43	8755.51	-1.600	8646.96	7641.43	13.159	1004	873	15.038
1960	8913.36	9080.16	-1.837	9705.07	8599.04	12.862	1089	947	14.993
1961	8542.22	8709.98	-1.926	9656.81	8565.86	12.736	1130	983	14.901
1962	8767.99	8961.38	-2.158	8914.79	7749.80	15.033	1013	865	17.599
1963	9065.19	9296.22	-2.485	10658.70	10124.30	5.278	1176	1089	7.981
1964	9280.00	9380.44	-1.071	10534.70	9658.54	9.071	1135	1030	10.232

1965	9415.20	9337.54	0.832	10536.20	9587.08	9.900	1119	1027	8.987
1966	9143.42	9147.53	-0.339	9634.76	9436.97	2.096	1054	1029	2.469
1967	9966.87	9957.81	0.091	11521.20	10642.10	8.261	1156	1069	8.167
1968	9877.43	9860.52	0.171	11651.20	10278.10	13.359	1180	1042	13.206
1969	10433.80	10433.00	0.007	12125.50	10824.70	12.017	1162	1038	11.995
1970	10038.20	10071.20	-0.328	11276.00	11306.80	-0.272	1123	1123	0.000
1971	9424.63	9461.32	-0.388	10068.20	10101.60	-0.331	1068	1068	0.000
1972	9750.04	9802.07	-0.531	10182.80	10221.50	-0.379	1044	1043	0.116
1973	10001.30	10072.80	-0.710	12019.80	12083.60	-0.528	1202	1200	0.198
1974	9877.51	9972.44	-0.952	11404.10	11491.40	-0.760	1155	1152	0.233
1975	10479.70	10606.20	-1.193	12981.00	13101.40	-0.919	1239	1235	0.303
1976	10041.40	10193.00	-1.487	12011.70	12153.90	-1.170	1196	1192	0.304
1977	10550.20	10461.90	0.844	14043.40	13626.30	3.061	1331	1302	2.191
1978	10785.80	10696.40	0.836	14156.20	13731.90	3.090	1312	1284	2.198
1979	10592.60	10886.20	-2.697	13563.20	13840.60	-2.004	1280	1271	0.677
1980	10900.20	11286.10	-3.419	14973.50	15381.70	-2.654	1374	1363	0.815
1981	10993.90	NA	-	14597.70	NA	-	1328	NA	-
1982	11105.30	NA	-	15310.80	NA	-	1379	NA	-

[a] Adapted from BBS (1976, pp.1–2, 4–10, 12–5, 26–9; 1979, pp.168–71; 1980a, pp.20–5, 30–1, 33–4, 36–7, 46–52; 1982, pp.232, 235–8, 240–1; 1984a, pp.39, 42; 1984b, pp.249–52, 255); BRRI (1977, p.89); World Bank (1982, Tables 2.5–6).

[b] Based on Boyce's (1985) hypothesis.

1978–9. This hardly justifies the difference between 'bumper' and 'poor' harvests. This raises some further doubts about the validity of the *revised series*.

Boyce's view about underestimation of yields in recent years relies on fragmentary evidence (Boyce, 1985, p.A38). Boyce relies on three sources of data to support his hypothesis of yield underestimation in the official agricultural statistics: (1) Farm level studies by Bangladesh Ministry of Agriculture and Forests (BMAF, 1981a); (2) Bangladesh Academy of Rural Development (BARD, see Rahim, 1977) and (3) Bangladesh Rice Research Institute (BRRI, see Clay *et al.*, 1978). BRRI studies indicate that the official yield of broadcast *aman* rice may be significantly lower than the actual yields. A BMAF study (BMAF, 1979) found that farmers' yields are far below the official estimates, which is contrary to the BRRI findings (see Pray, 1980, p.17). Pray, however, regards BRRI estimates to be more reliable as they are based on actual crop cuts as against farmers' assessment of their own yields reported in the BMAF study (Pray, 1980, p.17).

How well do the BMAF (1981a) yields represent the actual transplanted *aman* yields? On Boyce's own admission 'purposive selection of villages may have introduced upward bias' (Boyce, 1985, p.A39). To the extent that this is so, yield underestimation in the official series is likely to be overstated by Boyce's approach. The upward bias resulting from purposive selection of BMAF survey villages apart, one cannot completely rule out the possibility of further upward bias. As Boyce (p.A32) indicates, within the Rice and Jute Estimates Reconciliation Committee which prepares the final official figures for crop output and area, there are pressures for both high and low estimates: while a high output estimate is likely to reflect favourably upon the Ministry of Agriculture, a low estimate would probably strengthen the hand of the Food Ministry for negotiations with international food aid agencies. This casts some further doubt on the extent of discrepancy between the actual yields and those reported in the official statistics.

Thus while the Boyce study is substantial, it is doubtful if the revised series is a significant improvement upon the official agricultural statistics. In the view of the present study, the real contribution of Boyce (1985) lies not so much in deriving a revised estimate of agricultural crop output as in identifying the areas where deficiencies in Bangladesh's official agricultural statistics exist and warrant further improvements.

Do the official data really underestimate growth rates in recent years? Some light can be thrown on this aspect by comparing growth rates using both the series for the two sub-periods: 1949–64 and 1965–80 (i.e. 1949–50 to 1964–5 and 1965–6 to 1980–1), using both data sets on foodgrains as well as overall crop output growth based on the Boyce data. The results are set out in Table 12.2. Estimates based on the official data indicate a constant annual growth rate of 2.37 per cent in the earlier period compared to 2.18 per cent in the later period. The growth rates in yield per gross cropped hectare are respectively 1.71 and 1.35 per cent. The use of revised data, however, indicates a movement of growth rates in the opposite direction: they increase from 1.42 per cent in the earlier period to 2.87 per cent in the later, with the corresponding yield growth rates being 0.60 and 1.90 per cent. The official statistics show falling growth rates, whereas the revised figures indicate rising growth rates.

One should note the overall poor explanatory power of the trend equations in both periods for the official data. However, the explanatory power of the trend equations is much poorer in the earlier period when one uses the *revised series*. Furthermore, one obtains poor fits for overall crop output growth during the 1949–64 period using Boyce's data. How can one resolve the dilemma of which data series provide the more realistic estimates of growth rates? Even assuming that the official data do underestimate growth rates in the later period and overestimate in the earlier, what is the guarantee that the revised series produces realistic estimates of growth rates? Given the poor statistical quality of the trend equations for the 1949–64 period and the inconsistencies contained in the observations corresponding to some years in the later period, how much reliance can one place on the estimates based on the revised data series on foodgrains or on the ones derived from the Boyce indices of the aggregate crop output growth?

Furthermore, one could ask why 1949–64 is chosen as the first period. Is it only because it divides the entire time period into to sub-periods of equal length? Does that adequately justify the selection of 1964–5 as the terminal year of the first period, or are there any other *a priori* considerations for deciding on such a cut-off point? Boyce does not make this clear. If the aim was to choose the second period (1965–80) so as to coincide with the period of adoption of the new technology, it would probably have been more appropriate to have taken 1967–8 or 1968–9 or 1969–70 as cut-off points because data on HYVs and related techniques are available from 1967–8.

Table 12.2 Exponential growth rates in agricultural and foodgrain production using revised and official data series: Bangladesh, 1947–8 to 1982–3

Period	Area			Yield			Output		
	Growth rate (%)	R^2	t-value	Growth rate (%)	R^2	t-value	Growth rate (%)	R^2	t-value
(a) Official data (foodgrain only)									
Period 1									
1947–8 to 1964–5	0.74	0.53	4.27	1.34	0.47	3.80	2.07	0.59	4.75
1947–8 to 1966–7	0.80	0.63	5.59	1.28	0.52	4.44	2.08	0.65	5.80
1947–8 to 1968–9	0.95	0.72	7.20	1.35	0.62	5.68	2.31	0.74	7.49
1947–8 to 1969–70	1.04	0.75	7.94	1.35	0.65	6.21	2.39	0.77	8.38
1949–50 to 1964–5	0.66	0.40	3.04	1.71	0.57	4.31	2.37	0.59	4.51
Period 2									
1965–6 to 1982–3	0.90	0.72	6.35	1.39	0.69	6.03	2.29	0.73	6.60
1967–8 to 1982–3	0.79	0.62	4.73	1.46	0.65	5.15	2.25	0.67	5.28
1969–70 to 1982–3	0.93	0.62	4.44	1.89	0.76	6.20	2.82	0.74	5.81
1970–1 to 1982–3	1.20	0.82	7.20	2.16	0.82	7.11	3.36	0.84	7.73
1965–6 to 1980–1	0.84	0.61	4.71	1.35	0.60	4.60	2.18	0.64	4.98
(b) Revised data (foodgrain only)									
Period 1									
1949–50 to 1964–5	0.76	0.48	3.58	0.66	0.18	1.73	1.42	0.38	2.91
Period 2									
1965–6 to 1980–1	0.98	0.69	5.58	1.90	0.86	9.24	2.87	0.86	9.28
(c) Boyce data (all crops)									
Period 1									
1949–50 to 1964–5	0.50	0.32	2.57	0.77	0.29	2.42	1.27	0.43	3.25
Period 2									
1965–6 to 1980–1	0.64	0.47	3.54	1.54	0.84	8.54	2.18	0.79	7.40

Note: All estimates except those on all crops are based on data presented in Table 12.1. Growth rates of all crops are based on indices set out in Table 10 of Boyce (1985, p.A39).

The present study has altered the sub-periods to determine what impact this has on estimated growth rates. Consider the sub-periods 1947–64 and 1965–82. The results are presented in Table 12.2. Trend equations fitted to the official data yield a higher growth rate (2.29 per cent) in the later period compared to the earlier (2.07 per cent). Again, change the cut-off points. Consider three alternative sub-periods – 1947–66, 1947–68 and 1947–69 – and the corresponding second sub-periods 1967–82, 1969–82 and 1970–82. It is found that with changes in the cut-off points the official data produce consistently higher growth rates in the second sub-periods than in the first sub-periods. It can also be seen that with 1969–70 included as the terminal year of the first sub-period, the growth rate in production rises considerably (by 40 per cent, from 2.39 per cent to 3.36 per cent) in the second sub-period. This is because production level in 1969–70 is higher than that in 1970–1. Even when 1969–70 is included as the initial year in the second sub-period, the growth rate is significantly higher (by 22 per cent, from 2.31 per cent to 2.82 per cent) than that in the first sub-period. Growth rates and their variation are sensitive to the way in which the period under consideration is subdivided into sub-intervals.

No theoretical or *a priori* reasoning is provided by Boyce for the choice of an exponential function for estimating growth rates. An exponential function provides a constant rate of growth within a particular period. In reality, however, there seems to be no reason for expecting food production to grow at a constant rate during either period. Indeed if linear equations are applied to the *revised* data, one gets equally good (even slightly better) statistical fits for both the periods. A linear function implies constant absolute change over time and declining growth rates if production is rising. Whether in fact the growth rate is rising, falling or remaining more or less stationary over the years, can be considered by employing othe functional forms (see, for example, Reddy, 1978; Rudra, 1970). In a later paper, Boyce (1986) fits a kinked exponential model to the revised Bangladeshi crop output data. But the kink is arbitrarily assumed to occur at the midpoint of the series (i.e. 1964–5) and this model, like the exponential one, estimates constant growth rates for the two sub-periods.

At this stage it is doubtful whether the *revised series* following Boyce (1985) is qualitatively superior to the official data. Unless there is a decisive superiority, the use of the *revised series* in preference to the official one is difficult to justify. Thus, despite significant limitations of the official data, there is no other comprehensive source of agricultural statistics in Bangladesh. One has no choice but to use them, while acknowledging their significant inadequacies.

12.3   TRENDS IN FOODGRAIN PRODUCTION AND IMPORT

Table 12.3 presents Bangladeshi data on production of rice and wheat (RICEP and WHEATP), which is used as a proxy for foodgrain production, import of the grains (IMPORTF), off-take from government stock (OFFTAKEF), total per capita consumption, per capita consumption from domestic sources, as well as per capita domestic production (PERCAPITA, DOMPERCP and DMPRDPHD) and the percentage of *imported foodgrains as a proportion of total available foodgrains (IMPRAVL)*. In aggregate terms, domestic food production has more than doubled during the last three decades.

One gets a completely different picture if foodgrain availability per capita is considered. Total foodgrain available for consumption is defined as domestic production, less 10 per cent deduction for seed, feed and wastage, plus imports.[2] Overall per capita foodgrain available for consumption (PERCAPIT) has not increased on the whole. In fact, the domestic component of foodgrain availability per capita (DOMPERCP) shows a declining tendency, as does the domestic production per capita (DOMPRDPHD). This implies that *during the last three decades Bangladesh has been unable to produce sufficient food to maintain the per capita food consumption of its growing population*, and per capita food consumption has declined marginally.

For centuries the Bengal region was self-sufficient in food, but in the last few decades it has become a net importer of foodgrains. Its agricultural production is erratic because of considerable year-to-year climatic variability (BBS, 1985b, p.303) and fluctuations in its domestic food production are as a rule compensated for by variations in grain imports (IMPORTF), as can be seen from Table 12.3. From an average (based on three years) of 147 thousand tonnes of imports per year in the early 1950s, foodgrain imports reached an average of 1.4 million tonnes in the early 1980s. Measured in per capita terms, imports increased from 3.3 kg to 18.7 kg during the period. In some years imports have exceeded 2 million tonnes and have been well over 1.5 million tonnes in a number of years in the last two decades. Trends in the import intensity of Bangladeshi food consumption are highlighted by considering food imports as a percentage of total

Table 12.3 Domestic production, availability and import intensity of foodgrains: Bangladesh, 1950–1 to 1984–5

Year	RICEP	WHEATP	IMPORTF	OFFTAKEF	PROCUREF	AVLFOOD	PERCAPIT	DOMPERCP	IMPPRAVL	DMPRDPHD
1950	7460.7	20.4	144.28	193.049	66.043	6860.00	158.429	155.496	2.103	172.774
1951	7147.6	23.2	94.49	254.012	19.305	6688.43	150.640	145.354	1.413	161.505
1952	7452.8	24.4	289.57	243.851	15.241	6958.09	153.262	148.226	4.162	164.696
1953	8377.0	24.1	202.19	111.765	26.417	7646.34	164.085	162.253	2.644	180.281
1954	7711.0	26.6	50.80	81.284	127.006	6918.12	145.034	145.992	0.734	162.214
1955	6486.6	22.6	172.73	50.802	0	5909.08	117.946	116.932	2.923	129.924
1956	8315.8	23.8	599.47	124.974	0	7630.61	149.620	147.169	7.856	163.522
1957	7720.2	22.8	684.82	98.557	33.530	7033.73	135.004	133.756	9.736	148.618
1958	7032.5	25.5	473.48	169.680	33.530	6488.35	122.191	119.627	7.297	132.919
1959	8617.8	29.2	621.82	194.065	200.161	7776.20	143.472	143.585	7.996	159.539
1960	9672.2	32.9	709.20	204.225	24.385	8914.43	161.493	158.235	7.956	175.817
1961	9617.2	39.6	414.55	259.092	26.417	8923.79	157.386	153.283	4.645	170.314
1962	8869.7	45.1	1459.04	721.393	10.160	8734.55	150.078	137.858	16.704	153.175
1963	10623.7	35.0	1018.08	451.125	4.064	10039.90	168.172	160.684	10.140	178.538
1964	10500.0	34.6	350.54	753.906	13.209	10221.80	166.751	154.668	3.429	171.853
1965	10500.6	35.6	937.81	961.180	94.492	10349.30	167.735	153.688	9.062	170.765
1966	9575.4	59.3	1117.65	1101.390	3.048	9769.58	153.852	136.555	11.440	151.728
1967	11458.7	62.5	1035.35	659.414	22.353	11006.10	168.547	158.791	9.407	176.435
1968	11543.8	107.4	1136.96	1085.140	9.144	11562.10	172.055	156.043	9.833	173.381
1969	12004.7	120.8	1571.82	1375.730	6.096	12282.60	177.238	157.474	12.797	174.971
1970	11143.2	132.8	1164.39	1338.130	6.096	11480.40	161.696	142.935	10.142	158.817
1971	9931.2	136.9	1715.09	1762.840	10.160	10811.00	148.953	124.811	15.860	138.679
1972	10091.4	91.4	2870.33	2660.010	5.080	11819.50	159.077	123.345	24.285	137.050

continued on page 258

Table 12.3   continued

Year	RICEP	WHEATP	IMPORTF	OFFTAKEF	PROCUREF	AVLFOOD	PERCAPIT	DOMPERCP	IMPPRAVL	DMPRDPHD
1973	11909.1	110.7	1692.73	1754.710	72.139	12500.40	163.618	141.595	13.541	157.327
1974	11287.3	116.8	2599.05	1785.190	129.038	11919.80	152.819	131.586	21.804	146.206
1975	12762.6	218.5	1468.19	1694.770	355.616	13022.10	162.980	146.220	11.275	162.467
1976	11752.6	259.1	807.76	1473.270	324.119	11959.70	146.206	132.158	6.754	146.842
1977	13695.9	347.5	1634.82	2029.040	581.179	14086.90	168.303	151.004	11.605	167.783
1978	13662.4	493.8	1180.65	1814.660	360.696	14194.50	165.824	148.839	8.318	165.376
1979	12740.2	823.0	2871.35	2440.540	269.252	14378.20	163.947	139.189	19.970	154.655
1980	13881.2	1092.3	1078.03	1550.490	1043.480	13983.20	155.541	149.902	7.709	166.557
1981	13630.3	967.4	1245.67	2068.670	301.766	14904.80	162.717	143.427	8.358	159.364
1982	14215.5	1095.3	1870.54	1936.580	192.033	15524.30	165.858	147.219	12.049	163.577
1983	14508.1	1211.4	2133.70	2042.250	276.365	15913.40	166.285	147.832	13.408	164.258
1984	14662.9	1463.1	2616.32	2619.370	354.600	16778.20	169.135	146.304	15.594	162.560

Note: 1950 refers to 1950–1 (July 1950 to June 1951), etc. RICEP and WHEATP respectively refer to production of (cleaned) rice and wheat. IMPORTF is import of rice and wheat. OFFTAKEF is off-take of foodgrains from government stocks. PROCUREF is internal procurement of food. AVLFOOD is defined as production of food less 10 per cent for seed, feed and wastage, plus OFFTAKEF, less PROCUREF. The above-named variables are measured in thousands of tonnes. PERCAPIT, DOMPERCP and IMPRAVL are per capita availability of total and domestic foodgrains and imported foodgrains as percentage of available foodgrains from all sources. DMPRPHD refers to per capita domestic food production. PERCAPIT, DOMPERCP and DMPRPHD are measured in kilograms.

Source: Adapted from BBS (1976, pp.1–2, 4–10, 12–5, 26–9; 1978, p.379; 1979, pp.168–71; 1980a, pp.20–5, 30–1, 33–4, 36–7, 46–52; 1982, pp.232, 235–8, 240–1; 1984a, pp.39, 42; 1984b, pp.249–52, 255, 691; 1985b, p.842; 1986b, pp.29, 167); BRRI (1977, p.89); World Bank (1982, Tables 2.5–6); Alamgir and Berlage (1973, p.45); EPBS (1969, p.120–1); Alamgir (1980, p.221).

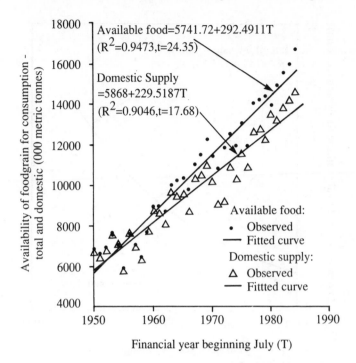

Figure 12.1   Trends in total available foodgrains (Available food) and
from domestic production (Domestic supply), Bangladesh, 1950–1 to
1984–5

available food for consumption (IMPPRAVL). This percentage increased from less than 2 per cent in the early 1950s to an average
exceeding 10 per cent in the last decade. In other words, *Bangladesh
now imports more than 10 per cent of its food requirements.*

Figure 12.1 illustrates the growing imbalance between Bangladeshi
demand for food and its supply from domestic sources, and Figure
12.2 shows the increasing import intensity of foodgrain supplies.
Trends in per capita food consumption (PERCAPIT), per capita
food consumption from domestic sources (DOMPERCP), and per
capita domestic food production (DMPRDPHD) are indicated by the
following linear regression equations:

$$\text{PERCAPIT} = 146.8321 + 0.5997T$$
$$(R^2 = 0.2215, \text{ } t\text{-value} = 3.060) \tag{12.1}$$

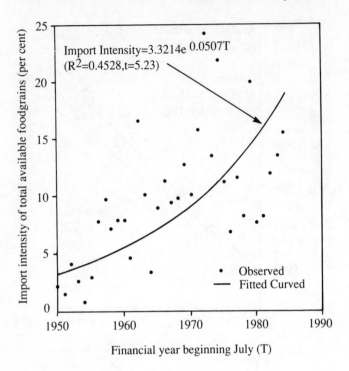

Figure 12.2   Trends in import intensity of foodgrains (Import intensity) as percentage of total available foodgrains for consumption, Bangladesh, 1950–1 to 1984–5

$$\text{DOMPERCP} = 145.7417 - 0.0654T$$
$$(R^2 = 0.0033, \text{ } t\text{-value} = 0.330) \qquad (12.2)$$

$$\text{DMPRDPHD} = 161.9352 - 0.0727T$$
$$(R^2 = 0.0033, \text{ } t\text{-value} = 0.330) \qquad (12.3)$$

The statistical fits of these regression lines are, however, poor. To the extent that any inferences can be drawn from them, they suggest that per capita foodgrain consumption shows a very slight upward trend. On the other hand, the domestic per capita food production (and availability of food per capita for consumption from domestic sources) possibly show a very slight downward trend. *To the extent that increased per capita food consumption has been achieved in Bangladesh, it has been achieved by increased dependence on im-*

ported foodgrains. It should be emphasized, however, that high import intensity of foodgrains does not necessarily imply non-sustainability. Japan imports a sizeable fraction of its foodstuff. However, Japan's balance of payments position is fundamentally different from Bangladesh's. While the former can pay for its food imports from export of manufactured goods, the latter has virtually no scope to do this. Bangladesh's exports consist, in the main, of agricultural commodities, primarily jute and jute goods, even though remittances by Bangladeshi workers abroad (especially from the Middle East) have added significantly to its export earnings in recent years (BBS, 1986b). Jute and jute products face keen competition from synthetics, while declining oil prices and the consequent recession is constraining labour exports to the Middle East from all countries including Bangladesh. It is difficult for Bangladesh to sustain its exports (cf. Tisdell and Fairbairn, 1984).

12.4   CONCLUDING COMMENTS

While the Green Revolution in Bangladesh has greatly expanded food supply, per capita gains are being thwarted by population increases, even though the latest census report shows some decline in the growth rate of population. Given the population growth rate and food growth rates of recent years, an increasing problem seems to be emerging. Population is increasing at 2.32 per cent per annum (BBS, 1985b, p.71) and, if the official statistics are used, a growth rate in foodgrain production of 2.18 per cent applied for the 1965–6 to 1980–1 period. So population is growing at a faster rate than food-grain production and, as suggested earlier, the gap is being met by growing import of grains. While the *revised series* gives a more optimistic picture of foodgrain growth rates (2.87 per cent) for the corresponding period, there are doubts about the reliability of the *revised series* based on the Boyce hypothesis. Agricultural production as a whole (that is, including foodgrains and other crops) is failing to keep pace with population growth, whether the official growth rate for 1965–80 is used (1.52 per cent) or Boyce's revised estimate of 2.18 per cent is adopted. Both sets of data indicate that foodgrain production is growing at a faster rate than agricultural production as a whole and suggest that cereal production is expanding at the expense of the non-cereal component. The latter effect is in fact confirmed by the results reported in Chapter 5.

We have observed that Bangladesh has become more dependent on food imports following the Green Revolution which, with its focus on cereal crops (which are readily transportable and storable), has encouraged increased urbanization. Food imports tend to be limited to urban areas because of the easier distribution of food to such areas and because of the potential influence of the urban population, which may become adjusted to the new diet and politically intolerant of temporary shortages of food.

Furthermore, in reality Bangladesh's food-import dependence is much higher than demonstrated above in terms of its direct dependence on imported foodgrains. Account must be taken of the import of inputs required for domestic food production, for example, chemical fertilizers, irrigation machinery, agro-chemicals and so on, as well as the import intensity of inputs critical to domestic production of some of the agricultural inputs, such as nitrogenous fertilizer, which have become increasingly used since the Green Revolution. These factors and other issues related to sustainability are considered in the next chapter.

13 New Agricultural Technology and Sustainable Food Production in Bangladesh

13.1 INTRODUCTION

Domestic food production has increased tremendously in Bangladesh as a result of its adoption of the new agricultural technologies associated with the Green Revolution. Unfortunately, however, Lester Brown's comments that such technology 'is literally helping to fill hundreds of millions of rice bowls once only half full' (Brown, 1970) does not apply to Bangladesh (cf. also Remenyi, 1988). In fact its expanding food production has not kept pace with population growth and Bangladesh has become increasingly dependent on imported foodgrain. More rice bowls are now half-filled or not quite half-filled! In addition, indications are that the growth rate of food production in Bangladesh is tapering off and that sustaining the growth rates of recent years is becoming ecologically more difficult. Furthermore, as a result of its change in agricultural technology (to higher 'tech' production), Bangladesh has become more dependent on foreign technology and imports of inputs required to maintain agricultural production. This dependence could also threaten the sustainability of Bangladesh's economic growth. While the present study does not wish to take a pessimistic view, there is cause for concern. In terms of the above general background, the main focus of this chapter is on the prospects for Bangladesh's sustainability of foodgrain production in relation to the new agricultural technology and the identification of factors which could influence this. This leads to a consideration of whether Bangladesh's food production is ecologically sustainable given (1) its increasing dependence on new ('modern') agricultural technology, (2) its reliance of foreign technology, and (3) its dependence on imports of inputs required for 'modern agriculture'. Bangladesh's situation is not unique in the world. Its food sustainability problem, for example, has parallels with those of some African countries and this makes its case of additional interest.

13.2   SUSTAINABLE DEVELOPMENT: THE ISSUES

Recently the question of sustainability of food production, given technological change, has received renewed attention (Douglass, 1984; Redclift, 1987).[1] Sustainability of per capita food production preoccupied classical economists such as Malthus (1798) and Ricardo (1817). While the Ricardian issues are still of concern to many (see, for example, FAO, 1984; IUCN, 1980), the focus has shifted to some extent to considering the sustainability of the ecosystems and environmental factors on which agriculture depends. Numerous institutions and individuals now argue that sustainable economic development is desirable, e.g. WCED, 1987; Barbier, 1987). This sustainability is likely to require, among others things, (1) sustainable ecological systems on which economic production ultimately relies (IUCN, 1980; Thibodeau and Field, 1984); (2) economic exchange that can be sustained if the country engages in international trade to a significant extent (Tisdell and Fairbairn, 1984); and (3) a socio-political structure that does not have within it the seeds of its own collapse (cf. Douglass, 1984). Thus the policy focus for economic development has shifted to some extent to stressing the importance of sustainability of ecosystems and the socio-economic and environmental factors on which agriculture relies (Turner, 1988; Wynen, 1989; Tisdell, 1990b).

Some economists, as opposed to some ecologists, see no particular virtue in a sustainable economic system or what in many cases amounts to a steady-state or stationary equilibrium for an economy. On the basis of the discounted present value criterion used by many economists (for example, Krutilla and Fisher, 1975), a productive system that is not sustainable may be preferred to the one that is. The issues are complex and are not debated here (Tisdell, 1983a; 1985a; 1988; Chisholm, 1988; Ruttan, 1988).

Given the interest of the present study in the sustainability of food production, the production characteristics of agricultural systems are of particular interest. Gordon Conway (1986) suggests that alternative agricultural techniques should be assessed on the basis of their influences on (1) the level of yields (**productivity**); (2) on the variability of yields and incomes (**stability**); (3) the *sustainability* of agricultural production (**sustainability**); and (4) the distribution of income (**equitability**). This is illustrated in Figure 13.1. A technique which results in higher yields, less variability of incomes, sustainable production and a more equal distribution of income is more desirable than the one without such characteristics. In practie, however, one technique rarely dominates another in terms of all of these characteristics. In such cases, difficult social choices and conflicts arise. Nevertheless, despite the limitations of Conway's test (Tisdell, 1985b), the characteristics mentioned by him are relevant in evaluating agricultural techniques. It is interesting that he includes sustainability as an important consideration, even though the concept of sustainability allows several interpretations (Tisdell, 1988).

Douglass (1984) pays particular attention to alternative meanings of agricultural sustainability. He claims that there are three different concepts in current use, as follows:

1. *Sustainability as food self-sufficiency*: The food self-sufficiency concept is a rather mechanical one. If a nation can meet current or future projected demands for food from its own agricultural resources, its demands are considered to be sustainable. Tests for this type of sustainability often hinge on whether or not a country has a net dependence on imported food.

2. *Sustainability as stewardship*: The stewardship concept of sustainability appears to have a number of strands to it. One view seems to be that it is unwise to push yields using non-renewable resources such as artificial fertilizers beyond levels that can be maintained by the use of renewable resources alone. For one thing, the use of non-renewable resources may damage or deplete the renewable resource-base, e.g. life-support system, so that future production is irreversibly restricted. Or another possibility is that 'too rapid application of non-renewable resources may cause a society to raise population and consumption levels unsustainably, leading eventually to a crisis breakdown of natural systems as they attempt to maintain the living levels to which they are accustomed' (Douglass, 1984, p.13).

3. *Sustainability as community*: The sustainability of community concept is also complex but is espoused by some 'alternative lifestyle' groups. It encompasses views on man's duty towards nature, the desirability of a fair distribution of income and full participation in the social and political systems of the community. Cooperation with nature rather than the domination of it is

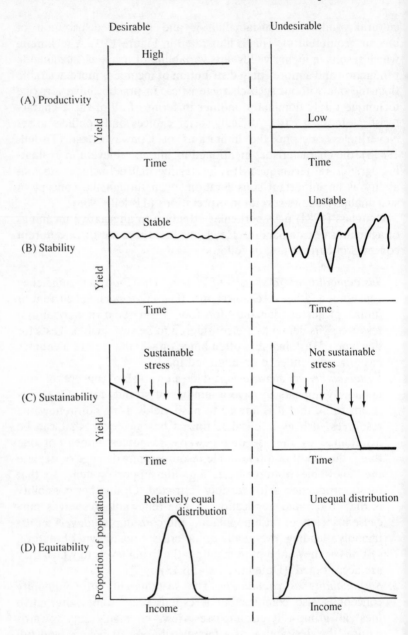

Figure 13.1   Desirable and undesirable properties of an agricultural ecosystem, according to Conway's criteria (Conway, 1985)

considered most desirable. Social structures which take power away from local farming communities and impersonalize them (such as capitalistic farming companies with headquarters distant from a farming community) are seen as threat to the community and to a sustainable economic order in harmony with nature (cf. Tisdell, 1983c).

Douglass says that

reserving some of their harshest judgements for modern agricultural technology for its spread to the Third World countries, alternative agriculturalists and others have developed evidence that the patterns of land tenure, market structure, and government policies which are found in many developing countries favour large land owners at the expense of peasants in the harvest of benefits from the new seed-fertilizer innovations, and this fact, they contend, is yet again destabilising to rural communities (Douglass, 1984, pp.19–20).

13.3 TO WHAT EXTENT ARE GROWTH RATES IN FOOD PRODUCTION LIKELY TO BE SUSTAINED?

A crucial question that arises at this stage is: Are the main trends continuing as in the past? Or is the Green Revolution tending to run out of steam? *While one must be wary of extrapolating past patterns*, it might be of interest to see if there is any statistical indication that growth trends in intensity of cultivation (INTENSITY), and yields per gross and per net cropped hectare (GYFDT and NYFDT) are tending to slow down. The data are set out in Table 13.1

To capture this aspect, logistic functions and, for comparison, linear functions, were fitted for each of these variables for the period 1947–8 to 1984–5. The results are presented as Equations (13.4)–(13.9). These are plotted in Figures 13.2 and 13.3. The linear functions give better fits in terms of the statistical quality of estimates even though the logistic functions also give a reasonable statistical fit. Because the linear functions indicate constant annual absolute change, they *imply declining growth rates*, given positive coefficients for the independent variables. The logistic functions suggest the view that the yields per net and gross cropped hectare are approaching upper limits of 2300 kg and 1500 kg respectively. The intensity of

Table 13.1 Net and gross cultivated area, net and gross land–man ratios, foodgrain yield per gross and net cropped hectare: Bangladesh, 1947–8 to 1984–5

YEAR	NCA	GCA	NCAPHD	GCAPHD	INTENSITY	GYFDT	NYFDT
1947	7.862	10.236	–	–	130.188	888.47	1156.68
1948	7.953	10.417	–	–	131.009	989.40	1296.20
1949	8.133	10.384	–	–	127.687	942.00	1203.77
1950	8.313	10.623	0.192	0.245	127.788	919.73	1175.31
1951	8.375	10.836	0.189	0.244	129.385	868.77	1124.05
1952	8.461	11.122	0.186	0.245	131.450	885.09	1163.45
1953	8.453	11.249	0.181	0.241	133.077	938.99	1249.57
1954	8.481	11.141	0.178	0.234	131.364	891.84	1171.56
1955	8.277	10.506	0.165	0.210	126.930	821.45	1042.67
1956	8.278	10.491	0.162	0.206	126.734	1020.76	1293.64
1957	8.218	10.492	0.158	0.201	127.671	940.60	1200.87
1958	8.044	10.216	0.151	0.192	127.001	883.46	1122.01
1959	8.326	10.715	0.154	0.198	128.693	1003.71	1291.71
1960	8.442	11.166	0.153	0.202	132.267	1088.87	1440.22
1961	8.475	11.086	0.149	0.196	130.808	1130.51	1478.80
1962	8.457	11.208	0.145	0.193	132.529	1016.74	1347.48
1963	8.532	11.403	0.143	0.191	133.650	1175.81	1571.47
1964	8.540	11.548	0.139	0.188	135.222	1135.19	1535.04
1965	8.742	11.955	0.142	0.194	136.754	1152.13	1575.57
1966	8.544	11.752	0.135	0.185	137.547	1053.78	1449.44
1967	8.801	12.723	0.135	0.195	144.563	1155.93	1671.06

Year	NCA	GCA	NCAPHD	GCAPHD	INTENSITY	GYFDT	NYFDT
1968	8.751	12.600	0.130	0.188	143.984	1179.63	1698.47
1969	8.807	13.290	0.127	0.192	150.903	1162.11	1753.66
1970	8.644	12.292	0.122	0.173	142.203	1123.33	1597.41
1971	8.244	11.400	0.114	0.157	138.282	1068.23	1477.18
1972	8.434	11.752	0.114	0.158	139.341	1044.39	1455.26
1973	8.489	11.907	0.111	0.156	140.264	1201.86	1685.77
1974	8.320	11.589	0.107	0.149	139.291	1154.49	1608.11
1975	8.485	12.013	0.106	0.150	141.579	1238.65	1753.68
1976	8.274	11.727	0.101	0.143	141.733	1196.27	1695.50
1977	8.374	12.623	0.100	0.151	150.740	1331.13	2006.55
1978	8.418	12.888	0.098	0.151	153.100	1312.46	2009.38
1979	8.447	12.940	0.096	0.148	153.190	1280.51	1961.63
1980	8.562	13.160	0.095	0.146	153.702	1373.72	2111.43
1981	8.584	13.208	0.094	0.144	153.868	1327.91	2043.22
1982	8.610	13.316	0.092	0.142	154.657	1378.73	2132.31
1983	8.651	13.250	0.090	0.138	153.157	1420.02	2174.31
1984	8.641	13.126	0.088	0.135	151.904	1477.06	2243.62

Note: 1950 means 1950–1 (July 1950 to June 1951), etc. NCA and GCA refer to millions of hectares of net and gross cropped area respectively. NCAPHD and GCAPHD respectively refer to hectares of net and gross cropped area per head of population. INTENSITY (cropping intensity)=(GCA/NCA) × 100. GYFDT and NYFDT are kg of foodgrain yield per hectare, gross and net cultivated area under foodgrains respectively. NYFDT = [GYFDT × INTENSITY]/100.

Source: Based on data from Alamgir and Berlage (1974, pp.161–7, 172); BBS (1979, pp.166–7, 340; 1984a, p.31; 1984b, p.570, 690; 1985b, p.301, 310; 1986a, pp.39, 47–50, 53; 1986d, pp.29, 167; EPBS (1969, pp.40–1, 120) and the sources mentioned in Table 12.1.

cropping seems to be hovering about the 155 per cent mark. This seems to be supported by the observations for the last few years.

$$\text{YFDT} = 845.8793 + 14.2697T$$

$(R^2 = 0.8409, t\text{-value} = 13.790)$ \hfill (13.4)

$$\text{YFDT} = 820 + (660) / (1 + e^{-(-2.7341 + 0.1289T)})$$

$(R^2 = 0.6194, t\text{-value} = 7.654)$ \hfill (13.5)

$$\text{NYFDT} = 1021.0986 + 28.6914T$$

$(R^2 = 0.8648, t\text{-value} = 15.170)$ \hfill (13.6)

$$\text{NYFDT} = 1042 + (1208) / (1 + e^{-(-3.2694 + 0.1500T)})$$

$(R^2 = 0.5868, t\text{-value} = 7.150)$ \hfill (13.7)

$$\text{INTENSITY} = 124.3189 + 0.7683T$$

$(R^2 = 0.8123, t\text{-value} = 12.480)$ \hfill (13.8)

$$\text{INTENSITY} = 126.5 + (29.5) / (1 + e^{-(-3.5268 + 0.1573T)})$$

$(R^2 = 0.7389, t\text{-value} = 13.093)$ \hfill (13.9)

These statistical results suggest that growth in availability of food in Bangladesh appears to be tapering off. However, we re-emphasize, as discussed in the beginning of this section, that it may be misleading to extrapolate past results. Growth rates of Bangladeshi food production are not being sustained despite the increasing use of the new agricultural technology. As discussed below, it may even prove to be difficult for Bangladesh to maintain some of the advances in foodgrain production that have been achieved in the past. It needs, however, to be emphasized that the findings presented in terms of Equations (13.4)–(13.9) are based on the official data and in view of the limitations discussed by Boyce (1985) and Pray (1980) must be qualified. Consider some factors that influence these trends.

Most of the growth in Bangladesh's food production in the last three decades has been achieved by the adoption of the new technology, associated in part with the introduction of high-yielding crop varieties. Biochemical technologies have played a dominant part in this growth (Diwan and Kallianpur, 1985; 1986). Opinions of scien-

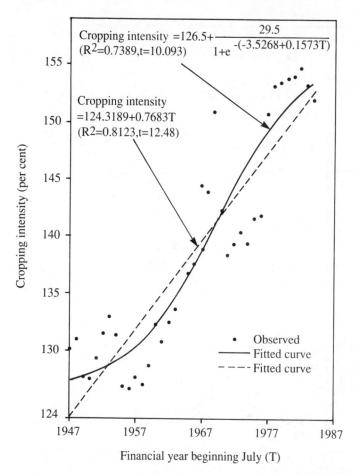

Cropping intensity $=126.5+\dfrac{29.5}{1+e^{-(-3.5268+0.1573T)}}$
($R^2=0.7389, t=10.093$)

Cropping intensity
$=124.3189+0.7683T$
($R^2=0.8123, t=12.48$)

- Observed
— Fitted curve
- - - Fitted curve

Financial year beginning July (T)

Figure 13.2   Trend in cropping intensity: Bangladesh, 1947–8 to 1984–5

tists and agricultural economists vary widely as to the extent that high levels of production based on such modern technology can be sustained. Conway (1986), relying on fragmentary evidence from LDCs, believes that on the whole modern agricultural technologies provide less sustainability for production than traditional methods. On the other hand, Schultz (1974), citing experience on the drylands of the United States, argues that modern technologies have led to agricultural systems that are environmentally more sustainable than in the past. It may be unwise, however, to generalize from a limited number of cases, for there are variations in modern technologies and in the

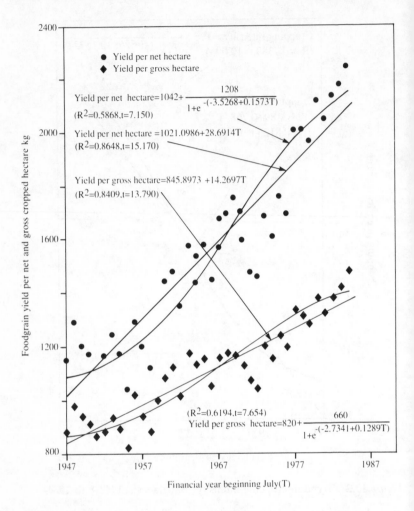

Figure 13.3   Trends in foodgrain yields per hectare of net and gross cropped area: Bangladesh, 1947–8 to 1984–5

environmental conditions under which they are applied, and so results may differ depending on particular conditions present (Tisdell, 1985b).

Yet it is clear that one cannot assume that high production levels which may be achieved by the use of modern technologies are sustainable. The possibility of such production being unsustainable deserves to be considered seriously. This requires one to be alert for signs of such non-sustainability and, where appropriate, to adopt scientific and social measures to counteract these trends or reduce their adverse impact on welfare.

A key element in the adoption of the new agricultural technology and the resulting agricultural output increase in Bangladesh is the increased intensity of cultivation. Bangladesh responded initially to population growth by extending cultivation and subsequently by intensifying it. This had a significant impact on the supply base of agriculture (see Alauddin and Tisdell, 1987; 1988b; see also Chapter 2).[2] However, the last five years have not witnessed any increase in cropping intensity. If anything it shows signs of decline, at least temporarily. One must, however, make allowance for the poor quality of data in this context. Despite a significant increase in the area triple cropped in the last two decades, it does not seem to have increased at all during the last five years (see Table 13.1).

A number of factors might have been at work. First, intensive cultivation of land during the *rabi* season is critically dependent on effective expansion of irrigation facilities. Yet the evidence of the last five years (see Table 13.1) indicates that the percentage of area irrigated for foodgrains (PRFDI) has not increased significantly. Secondly, increased cropping intensity is critically constrained by the maturity periods of crops. Moreover, over-intensification, e.g. triple cropping in a year, may reduce overall yields by creating deficiencies of important soil nutrients. This matter and related long-term effects, mentioned below, require continuing study by agricultural research bodies (cf. Alauddin and Tisdell, 1986a; Gill, 1983). Furthermore, soil structure can also be changed by increased frequency of cultivation. It may lose organic content (humus) and become more subject to erosion and lose its water absorption and aeration properties, all of which may reduce yields.

Now consider the extent of adoption of HYVs in Bangladesh, another important contributor to growth in agricultural production. After very rapid expansion during the initial years, the percentage of land under HYV foodgrains (PRFDHA) appears to be slowing down. One can see from Table 13.2 that is seems to be stabilizing around 30 per cent. It can also be seen that the per cent areas under *kharif* and *rabi* HYVs of foodgrains (PRKHA and PRRABIHA) appear to have stabilized around 17 and 80 respectively. While irrigation is the limiting factor for expansion of area under *boro* HYV, this may not be the only factor. In the low-lying areas of (greater) Mymensingh and Sylhet districts, local variety *boro* rice is cultivated primarily because *boro* HYVs suitable for these areas are not yet available to farmers.

Table 13.2   Intensity of modern input use: Bangladesh agriculture, 1967–8 to 1984–5

YEAR	PRKHA	PRRABIHA	PRFDHA	PRFDI	FPHA	IRKHP	IRRBP
1967	0.044	9.207	0.686	NA	7.898	NA	NA
1968	0.097	16.513	1.648	NA	8.499	NA	NA
1969	0.309	24.315	2.618	8.725	9.908	20.787	79.213
1970	1.268	32.566	4.720	10.174	11.771	15.866	84.134
1971	3.592	33.293	6.779	10.304	10.170	13.869	86.131
1972	7.223	41.780	11.140	10.863	15.459	13.839	86.161
1973	10.874	52.621	15.772	11.583	14.890	16.297	83.703
1974	9.088	55.483	14.946	13.103	11.371	14.173	85.827
1975	9.908	56.266	15.651	12.100	17.626	12.450	87.550
1976	8.729	59.952	13.904	10.474	20.343	15.294	84.706
1977	9.723	59.837	16.069	12.145	26.941	13.234	86.766
1978	11.861	64.331	18.694	12.225	25.804	14.419	85.581
1979	14.139	71.688	22.735	13.229	28.794	15.733	84.267
1980	15.818	75.254	25.368	13.334	31.955	17.897	82.103
1981	15.587	76.990	25.845	13.904	30.171	19.450	80.550
1982	16.921	80.850	28.161	14.885	37.945	19.333	80.667
1983	17.097	81.635	28.330	15.496	41.878	17.705	82.295
1984	17.874	83.819	31.492	17.113	46.158	15.910	84.090

Note: 1967 refers to 1967–8 (July 1967 to June 1968), etc. PRKHA and PRRABIHA respectively refer to percentages of *kharif* (rain-fed rice, *aus* and *aman* varieties) and *rabi* (dry season, *boro* rice and wheat) area under HYV cultivation. PRFDHA represents percentage area under HYV foodgrains. FPHA is fertilizer (kg of nutrients) applied per hectare of gross cropped area. IRKHP and IRRBP are percentages of *kharif* and *rabi* foodgrain area irrigated in total irrigated area. NA means not available.

Source: Based on data from BBS (1979, pp.162, 212; 1982, pp.206, 209, 213; 1984a, pp.31, 33; 1984b, p.225; 1985b, pp.301, 310; 1986a, pp.39, 41–3, 47–50, 53, 70, 72).

The possibility that modern agricultural systems are eroding or damaging *natural* environmental life-support systems for agriculture must also be considered.

First, increased use of chemical fertilizers necessary to maintain or increase yields can result in acidification of soils and change their desirable structures. Furthermore, not all fertilizer applied is used by crops, and can become a source of non-point pollution. Excess fertilizers may drain from the land into surface or ground water or be tied into the soils. Particularly in soils with inadequate humus, the efficiency of conversion of fertilizer to plant tissue is low, and the ability of soils to store reserves of nutrients is poor. Excess nutrients from agricultural land drainage disrupt normal ecological succession and promote blooms of blue-green algae. In the event of extreme pollution all aerobic organisms disappear, and only a few tolerant fish species may survive (M.F. Ahmed, 1986, p.44).

Secondly, irrigation can have an adverse impact on ecology and environment. It can result in saline soils (see, for example, Sinha, 1984, p.181). The withdrawal of ground water, in areas coupled with inadequate recharge of aquifer in the dry season, has lowered the ground water table beyond suction limit in many areas, especially in the northern part of Bangladesh, and has disrupted the drinking water supply system based on hand pump tubewells in those areas (M.F. Ahmed, 1986, pp.47–8).

Thirdly, pesticides, particularly insecticides, affect not only the target species but also non-target populations, e.g. predators and parasites which may prove beneficial to human beings Indiscriminate use of pesticides causes environmental pollution as the toxic chemicals are washed away by rain to water bodies. Fisheries are affected (Hamid *et al.*, 1978) and the residual toxicity may end up causing human health hazards. Other long-term environmental effects of pesticides cannot be fully predicted (M. Ahmed, 1986, p.91). The use of agro-chemicals, it is argued, can reduce species diversity, upset ecological balance and stimulate the development of pest populations. While according to Hayami and Ruttan (1985, p.297), extreme critics view the development of modern varieties 'as a plot by multinational firms and foundations to make peasant producers. . .dependent on chemical fertilizers and pest control materials', a more moderate view points to the dangers from loss of genetic diversity (this point is further taken up in the following section) and ecological balance. Multiple cropping can reduce humus levels and lower soil nutrients, and frequent cultivation can cause a deterioration in soil structure. There can, therefore, be a long-term decline in the *natural* fertility of the soil akin to a 'mining-effect'. Furthermore, agricultural production has become increasingly dependent on man-made chemicals, most of which are derived from non-renewable resources. This could create further problems as discussed in Section 13.6.

13.4   EVIDENCE FROM FIELD SURVEYS IN TWO VILLAGES

Now consider some farm-level evidence relating to sustainability of agricultural production. Some findings based on the data gathered in the course of the survey (for details see Chapter 7) are worth reporting. The following observations are based on Table 13.3 as well as other information collected during the fieldwork and may be relevant in identifying obstacles to achieving greater food production in Bangladesh.

Triple cropping of paddy, though possible if two short duration HYVs are grown in the *boro* and *aus*, is rarely practised, and farmers reported lower overall yields, possibly due to zinc or sulphur deficiency. However, to replenish the lost soil nutrient, some farmers reported using fertilizers containing zinc, on the advice of the agricultural research workers.[3] From the field survey it is clear that farmers in both survey areas of Bangladesh, *ceteris paribus*, are using more chemical fertilizers per hectare to maintain yield. Respondents were asked: 'Do you have to apply more fertilizer per hectare than before to maintain yield?'. All the farmers in the Comilla village said that they had to apply increasing amounts of fertilizers to maintain yield. In the Rajshahi village, 88 per cent of the farmers claimed that it was becoming necessary for them to apply more fertilizer in order to maintain yields. The farmers with a different opinion, with a solitary exception, were either non-adopters or did not have a long experience with HYV rice cultivation.

There is also independent (objective) evidence from the survey to support the hypothesis that increased quantities of application of chemical fertilizer are becoming necessary to maintain yield. Farmers were asked for information on fertilizer use and yields over a number of years. The average results are given in Table 13.3. As can be seen from the table whereas fertilizer use per hectare has risen quite significantly in the last five years, there was little evidence of increase in yields of HYV foodgrains overall. In Ekdala, fertilizer use per net hectare cropped with HYV rice rose by 44 per cent between 1980–1 and 1985–6, while the yield declined by 7 per cent. The HYV yields in South Rampur show some marginal increase, but only as the result of a substantial 88 per cent increase in fertilizer application during the same period. [4]

HYVs in the survey areas have been adopted on all acreage (or virtually all) agronomically suitable for them and having irrigation facilities (for further details, see Chapter 5). For instance, there are ecological constraints on expanding HYV cultivation of rice during the *rabi* as well as *kharif* seasons in South Rampur which is flood prone. In this village the percentage area under *aman* HYVs is just above 40 and there has been no tendency for this area to increase in the last few years. The reluctance of the farmers to adopt HYVs

Table 13.3   Selected indicators of technological change and productivity trends in the 1985–6 crop year and previous five years: evidence from farm-level data in two areas of Bangladesh

(a) 1985–86 crop year		
Variables	*Ekdala*	*South Rampur*
Area under irrigation (ha)	30.4	52.2
	(44.8)*	(100.0)*
Area under HYV cultivation (ha)		
Rabi season	32.2	52.2
	(47.4)*	(100.0)*
(Wheat)	(11.2)	
Kharif season	27.3	21.0
	(40.4)*	(41.0)*
Cropping intensity (per cent)	175.1	198.4
Fertilizer use (kg/net cropped ha of HYV rice)	849	557
HYV rice yield (kg/gross cropped ha)	3431	2596
(Rice yield/gross cropped ha)	(2796)	(2361)

(b) Previous five years									
	Ekdala					*South Rampur*			
	1980	*1981*	*1982*	*1983*	*1984*	*1980*	*1981*	*1982*	*1983*	*1984*
Area under irrigation (ha)	32.9	33.5	33.2	30.7	32.1	51.0	51.0	51.3	50.9	49.7
Area under HYV cultivation (ha)										
Rabi season	28.7	31.7	30.7	31.8	30.9	51.0	51.0	51.3	50.9	49.7
Wheat	6.4	8.4	9.7	11.2	11.0					
Kharif season	25.3	26.4	25.4	23.7	24.3	13.6	13.6	15.7	15.7	17.4
Fertilizer (kg/net ha of HYV rice)[a]	591	608	670	713	724	296	297	329	396	425
HYV rice yield (kg/gross ha)	3700	3651	3504	3551	3486	2558	2563	2662	2687	2762

* Per cent of total operated area.
[a] Average of 58 farmers in each area. These figures are in terms of gross weight of fertilizers and not in terms of nutrient contents (cf. Table 13.2).

beyond that limit stems from two considerations: (1) flood levels at times are beyond the tolerance of the currently available strains of *aman* rice; (2) in case of severe floods, the yields of HYVs fall far below those of their local counterparts. Therefore, as a hedge against uncertainty, farmers allocate land between local and HYVs of *aman* rice. On the other hand, the adoption rate for HYVs of foodgrains during the *rabi* season is 100 per cent, as the entire cultivable land in the village is under irrigation and therefore environmental conditions are controlled. In both cases, the adoption of available HYVs has reached its limit.

Now consider the situation in the drought-prone village of Ekdala in the district of (greater) Rajshahi. Overall HYV adoption is much lower in Ekdala (40 per cent during *kharif* season and 45 per cent during *rabi* season) than that in South Rampur. The area under HYVs has not increased in recent years in either season. The factors that help explain the pattern of adoption of HYVs in Ekdala are, among others: (*a*) inadequate irrigation facilities during the dry season; and (*b*) lack of drought-resistant qualities of the *aus* and *aman* HYVs of rice. In case of drought of a serious nature, yields of the HYVs fall far below those of the local varieties. Farmers having land under irrigation were generally found to allocate land to *aus* amd *aman* HYVs of rice as they can provide supplementary irrigation in case of insufficient or delayed rainfall during the *kharif* season. Farmers without land under irrigation do not allocate land to HYVs of rice in either season and normally cultivate other crops like sugar cane, wheat, pulses, etc. This is in sharp contrast to the cropping pattern of South Rampur where land is allocated to rice alone: *boro* rice during the *rabi* season, and local and HYV *aman* rice during the *kharif* season.

At a more general level, it has been suggested that modern agricultural technologies are reducing the available genetic base. According to Biggs (1982), as cited by Hayami and Ruttan (1985, p.297), in 1977 the wheat crop in the Sonora Valley, the home of the Mexican dwarf wheat varieties, was threatened by a large-scale outbreak of rust because of lack of genetic diversity in the available commercial varieties. Biggs and Clay (1981, p.330) cite another recent incident, namely, a significant reduction in the 1978 Pakistan wheat yields because of dependence on a set of rust-susceptible wheat varieties. Both incidents, as well as additional cases cited by Biggs and Clay (p.330), underscore problems posed by dependence on a narrow range of genetic materials. This can make it more difficult to sustain long-term production. The farm-level data from Ekdala indicate that farmers primarily rely on two or three varieties of HYV rice. During the 1985–6 crop year, BR11 and China varieties constituted 82 per cent of the gross area planted to HYV rice. In South Rampur, farmers were found to allocate over 70 per cent of HYV rice area to four rice varieties: BR3 (21 per cent), BR11 (26 per cent), *Paijam* (13 per cent) and Taipei (10 per cent).

Table 13.4   Imports of modern agricultural inputs: Bangladesh, 1973–4 to 1983–4

Year	Fertilizers ('000 tonnes of gross weight)				Chemicals (million taka current c.i.f)		Irrigation	
	Urea	TSP	MP	Others	Pesticides	DTW	STW	LLP
1973	0	97.13	41.25	10.87	105.50	289.10	7.10	110.80
1974	142.25	48.16	7.01	35.46	118.70	508.50	8.80	343.40
1975	72.34	222.72	37.39	0	21.90	266.70	44.50	0
1976	10.97	20.83	9.96	0	35.00	146.70	54.60	78.00
1977	260.31	114.61	37.49	0	67.80	203.40	13.10	27.20
1978	353.79	191.52	78.03	0	76.00	57.00	35.20	0
1979	286.32	215.81	60.05	11.28	80.60	NA	492.50	6.90
1980	62.99	224.55	42.67	19.31	152.80	162.40	171.30	83.20
1981	254.00	184.00	26.00	0	147.70	174.80	293.20	89.60
1982	43.00	208.00	44.00	9.00	119.14	NA	415.73	NA
1983	94.00	124.00	60.00	127.00	250.48	NA	331.89	NA

Notes: 1973 means 1973–4 (July 1973 to June 1974), etc. DTW, STW and LLP respectively represent deep and shallow tubewells and low lift pumps. NA means not available. Fertilizer imports are in terms of thousand tonnes. Import data on pesticides and DTW, STW and LLP are in current millions of taka in c.i.f. value (cost, insurance, freight).

Source: Based on data from BADC (1978, p.14; 1979, p.8; 1980, p.9; 1981, p.9; 1984, pp.1–2; 1985, p.7); BBS (1985a, pp.867–71; 1985c, pp.477–9).

13.5   DEPENDENCE ON FOREIGN TECHNOLOGY AND IMPORTED INPUTS

The use of modern agricultural technology by Bangladesh has made it increasingly dependent on foreign technology and know-how. Taking this into account, Bangladesh has become more food-dependent on foreigners than is suggested by its need to import an increasing proportion of its grain supplies (see previous chapter).

Modern seed-fertilizer-irrigation technology is highly import intensive. Bangladesh now depends heavily on the foreign supply of chemical fertilizers, pesticides and fungicides as well as irrigation equipment. Data set out in Table 13.4 clearly indicate the extent of import dependence of Bangladesh for the inputs critical to augmenting domestic food production. Bangladesh primarily produces two types of nitrogenous fertilizer, urea and ammonium sulphate (AS), and one type of phosphatic fertilizer, triple super phosphate (TSP).

Bangladesh depends heavily on external supply for other non-nitrogenous fertilizers like muriate of potash (MP), hyperphosphate (HP). The production of TSP is entirely based on imported rock phosphate and Bangladesh has not been able to secure a long-term source of supply of phosphate rock for TSP production (Alamgir, 1980, p.211), even though Bangladesh produces most of its nitrogenous fertilizer from domestic natural gas.

While in its production of nitrogenous fertilizers (urea and AS), Bangladesh does not import its basic raw materials (natural gas), it is very import dependent (at least indirectly) for several other materials and inputs required to manufacture nitrogenous fertilizer. An examination of Bangladesh's inter-industry input–output structure indicates that sectors such as other chemicals, transport services, etc., are important suppliers of inputs to the fertilizer sector, and these sectors are heavily dependent on imported inputs like petroleum, machinery, chemicals and so on for the production of their output (see, for example, BPC, 1980, p.50–3).

Recent studies (Alauddin and Tisdell, 1986b; 1988a) show that, along with many other industries in Bangladesh, fertilizer production involves higher capital–labour ratios, in terms of techniques used, than seem appropriate given the total factor composition of the economy. It is possible that Bangladesh's dependence on import of capital equipment and/or the dependence on foreign technology (required for industrial manufacture of inputs necessary for agriculture) will continue to grow if it tries to augment domestic food production (cf. Maitra, 1980; 1988).[5]

13.6   CONCLUDING OBSERVATIONS

It seems that Bangladesh's food supply problem will remain critical into the foreseeable future. The crucial question is whether or not the recent growth rates can be sustained. A recent study by FAO (1984, p.28) on food supplies observes, 'Bangladesh would be critical at intermediate input levels, and even with high inputs would be able to support only 16 per cent more than its expected year 2000 population, although it is projected to grow by another 50 per cent by year 2025'.[6] The FAO study suffers from two serious limitations that may lead to underestimation of the gravity of the food supply problem in Bangladesh. First, high input levels (approximately equivalent to the

Wetern European level) are clearly unattainable because of the impossibility of generating sufficient funds to pay for imports of required inputs. For Bangladesh, a large part of modern agricultural inputs is imported and even those that are domestically produced are highly import intensive. Secondly, it assumes that the 'whole potentially cultivable land is used to grow nothing but food crops. . .No allowance is made for other essentials such as fibres or vegetables and fruits. . .' (p.10). It therefore substantially reduces, if not virtually eliminates, Bangladesh's capacity to earn enough foreign exchange from fibre sales (e.g. from jute sales). Foreign exchange is also required to sustain imports of agricultural inputs for domestic production and to obtain other consumption, intermediate and capital goods.

The restoration of Bangladesh's food supply–demand balance depends critically on a significant reduction in population growth and an increase in yield per hectare. While the agronomic potential for increasing yield is substantial, the realization of this potential to reality requires increased intensity of cropping. But, as mentioned earlier, there may be ecological limits to the degree of intensification that is sustainable. Also account needs to be taken of major socio-economic constraints on achievement of agronomic potential. According to Boyce (1987a, p.322) 'the key constraint on agricultural growth in Bangladesh is the inability of cultivators to achieve water control – that is, irrigation, drainage and flood control – within the current framework of a highly disarticulated and inequitable agrarian structure'.

In a wider context, Bangladesh, along with other countries, may encounter problems in the long run in sustaining agricultural productivity because of loss of genetic diversity, ecological damage due to application of agro-chemicals and from its dependence on chemical products, for example, fertilizers and pesticides produced from exhaustible and non-renewable resources. Another difficulty for Bangladesh is that increased food production using modern techniques tends to raise its dependence on imported technology and imported inputs.

It is uncertian whether the new agricultural technology introduced into Bangladesh will result in a sustainable higher production per capita than in the past. The present chapter has attempted to identify obstacles to such sustainability. However, the conclusions of this chapter must be qualified for three reasons. First, the quality of the official data needs to borne in mind. Secondly, it would be unwise to generalize from a limited number of observations from only two villages in Bangladesh. Thirdly, it is conceivable that there may be major breakthroughs in new agricultural technology in the future. Nevertheless, continuing concern about the issues involved is not misplaced.

14 Concluding Overview and Comments

The present study considered the extent to which the 'Green Revolution' has been effective in Bangladesh; by (1) increasing the growth rates of output and yield; (2) ameliorating inequality and poverty; (3) moderating fluctuations in production and yield; and (4) providing a sustainable basis of agricultural production. In examining these social and economic aspects of technological change in Bangladeshi agriculture, the present study employed both primary and secondary data.

A number of findings of this study are consistent with those suggested by previous studies, whereas others differ significantly from those in the existing literature. The study yielded several important conclusions which can be summarized as follows:

Major changes have taken place in Bangladeshi agriculture since the introduction of the new agricultural technology. These changes include, among other things, the increase in the incidence of multiple cropping and decrease in net cultivated area, changes in cropping pattern and output-mix. Higher growth rates in output and yield of foodgrains have followed the Green Revolution, typifying a shift in Bangladeshi agriculture from relatively extensive methods to intensive ones. The overall rate of growth of crop output has not been very impressive even though total food production has increased significantly. This increase has occurred primarily through a moderate rise in the output of rice and a spectacular boost in the production of wheat. The period of the new technology also witnessed a trend away from non-cereal to cereal production.

We demonstrated that the decomposition methods such as that developed by Minhas and Vaidyanathan (1965) and applied to Bangladeshi data by such writers as Wennergren *et al.* (1984) are likely to give a misleading picture of productivity effects of technological change. When the area planted to a particular crop or variety of a crop increases, and the above decomposition methods are applied, there is a tendency to under-represent the yield effect if some fall in the average yield per hectare occurs and to over-represent the area effect. Since the Green Revolution has increased the area under several varieties of crops (for example, dry season crops in Bangladesh), the yield effect often appears to be negative or underestimated by these methods. This can occur even when the productivity of the crop on *every* unit of land is raised. To say that yield decreases in such a case because *average* yield per hectare per crop variety falls is not only ingenuous but grossly misleading.

Decomposition methods of the Minhas and Vaidyanathan type mask an important effect of the Green Revolution on production. One of its most important impacts has been to increase the intensity of cultivation or cropping by increasing the incidence of multiple cropping, that is, the number of crops grown on cultivated land in each year. Decomposition methods take account of this as an increase in area devoted to particular crops. Even if yields should fall during the original cropping period, the extra cropping in each year usually much more than offsets any fall during the original cropping period. This change reflects an important way in which the new varieties have added significantly to agricultural production. A serious shortcoming of the decomposition methods considered in this study is their failure to identify such effects.

The major contribution of the Green Revolution to increased agricultural output in Bangladesh seems to lie more in its contribution to increased productivity of already cultivated land through multiple cropping rather than in its contribution to extension of the area of cultivated land or to greater yields from single cropping. The present research has provided a more rigorous and in-depth analysis of the underlying sources of growth than some previous studies (e.g. M. Hossain, 1984; Pray, 1979; Wennergren *et al.*, 1984) which employed the decomposition methods.

In Chapter 3, the D-K model (Diwan and Kallianpur, 1985) was applied to Bangladeshi data for the first time. While the results of the application of the D-K model for Bangladesh seem more plausible *a priori* than the D-K estimates for India, estimated coefficients measuring the contribution of biological technology are low. Diwan and Kallianpur found lower γ coefficients for India than they expected *a priori*. At first sight, these results suggest that biological inputs have played a much smaller role in increasing food productivity, and in particular innovations of the Green Revolution have been less significant in this regard than has been commonly supposed. While this conclusion may be warranted, there are grounds for scepticism. In particular, important problems of identification arise when the D-K method is used, and aggregation problems need to be overcome. Other studies (M. Hossain, 1985; Hayami and Ruttan, 1984) do not support the D-K view.

Employing district-level data, it was found in Chapter 4 that the picture of overall Bangladeshi foodgrain output growth conceals important inter-district differences in growth rates of production and yield. There is evidence that the inter-district divergence in growth rates might have been moderated to some extent in the post-Green Revolution period. These aspects are not apparent from previous studies of Bangladeshi agriculture (Mahabub Hossain, 1980; Chaudhury, 1981b; Alauddin and Mujeri, 1986a) which compare inter-district growth rates of foodgrain production and yield in the period broadly consistent with that of the new technology.

Considerable differences in the pattern of use of modern inputs (irrigation, fertilizer, HYVs) have led to regional variations in output and productivity growth. Different regional patterns in the spread of the new technology have resulted in similar unevenness in intensity of cropping. While both irrigation and chemical fertilizers have significant positive impact in increasing foodgrain yields, percentage area under HYV emerged as the most dominant determinant of the increase in overall output per hectare. The availability of irrigation and the area potentially suitable for HYV *aus* and *aman* crops have a significant positive impact on expansion of HYV area.

The socio-economic variables, namely the proportion of farmland consisting of small farms and farmland under share tenancy, appear to have considerable influence on the intensity of adoption of HYVs. Concentration of land ownership and control appears to be an obstacle to the diffusion of HYV technology. As for the incidence of sharecropping, its impact varies according to the season. For instance, it has a negative impact during the (irrigated) *rabi* season but a positive impact during the (rainfed) *kharif* season. Overall, however, percentage area under HYVs seems to be positively associated with percentage of land under sharecropping. For this reason, the variables used as technological proxies and those that are surrogates for agrarian structure should be considered as important determinants of the degree of adoption and diffusion of the Green Revolution technology.

The above finding contrasts with that of M. Hossain (1980) who uses 1960 data for sharecropping and late 1970s data for percentage of area under HYV. Furthermore, seasonal adoption of HYVs is not explicitly considered by Hossain.

Chapter 2 identified the phenomenon of comparative 'crowding out' of non-cereals following the Green Revolution in Bangladesh, a trend similar to that observed for instance by Sawant (1983). Growth in production of pulses, fruits, spices, and to some extent vegetables, has failed to keep pace with that of cereals. On the whole, increased cereal production has been accompanied by rising population, with per capita cereal availability remaining roughly constant However, availability per capita of pulses, fruits and spices has fallen markedly following the Green Revolution, and on average the availability of vegetables per head has fallen. In addition, per capita protein content of the Bangladeshi diet has fallen markedly. The average Bangladeshi diet now appears to be less varied and balanced, and therefore less nutritious, with adverse welfare implications. While more individuals are being supported, they seem to be supported at a lower level of individual welfare on average, a feature which is not highlighted by the approach of Hayami and Herdt (1977).

Using a partial equilibrium analysis of Hayami and Herdt (1977), it was found that consumers' surplus is much greater than it would have been if the HYVs had not been introduced. By keeping the real price lower than otherwise, the modern varieties of cereals have tended to be income-equalizing for urban consumers. However, the impact of technological changes on incomes of farmers, and the distribution of income between those involved in production, is more complex.

Given the relatively inelastic demand for rice in Bangladesh, it seems that the real cash income of producers has risen slightly as a result of the new technologies. However, if a less partial view is taken and if account is taken of the lower cost of obtaining home-consumed produce, the increase in income may be still greater.

It is pointed out that the adoption of HYVs has led to important variations in factor shares in Bangladeshi rice production. For instance, the relative share of labour has fallen. However, the absolute share has increased, and it seems that rural employment has risen at least slightly as result of the new technologies. However, with population increase, unemployment and underemployment may still rise.

Using farm-level data from two Bangladeshi villages and bivariate and multivariate techniques, the influences of such variables as subsistence pressure, tenancy, labour scarcity, education, and availability of irrigation on the adoption of the HYV technology were identified in Chapter 7. It was found that the degree of access to irrigation is a key determinant of HYV adoption both within and between the villages. This is broadly consistent with the findings of Chapter 4.

Significant inter-village differences exist in the adoption of technology during the dry season. However, little difference exists in the extent of adoption of HYVs during the rainy season. In South Rampur, since everyone irrigates, the significance of irrigation does not show up from statistical analysis.

The adoption pattern clearly indicates that the currently available HYVs are neither sufficiently flood-resistant nor drought-resistant. Under the present circumstances, it is unlikely that the area under HYVs in the rainy season will increase much further.

We found empirical support for the hypothesis that the *intensity of adoption* of HYVs tends to fall with farm size. This seems to support I. Ahmed's (1981) and Asaduzzaman's (1979) hypothesis (postulating an inverse relationship between firm size and intensity of adoption) but not Jones's (1984) U-shaped relationship.

As for the dynamics of adoption (*the crude adoption rate* over time), there is support for the hypothesis that larger farms are earlier adopters but that smaller farms rapidly catch up with larger farms. This phenomenon seems to be consistent with the views and findings of Herdt (1987), Muthia (1971) and Ruttan (1977), but not with those of Hayami and Ruttan (1984) and Barker and Herdt (1978).

Empirical analysis of the characteristics of innovators and imitators (i.e. those adopting before and after the dividing line of half the average length of adoption) in Ekdala supports the above hypothesis. However, no firm conclusion can be drawn from the South Rampur data. The involvement and proximity of South Rampur to official crop experiments, trends in subsidization of inputs and frequent visits by the members of the Bangladesh Academy for Rural Development (BARD) may, however, have made it atypical.

Even though innovators and imitators alike have contacts with extension agents or have access to formal channels of information, it is the demonstrated success (e.g. seeing others' success, i.e. informal channels of information), that seems to have had the greater impact on farmers' decisions to adopt new technology. Even the innovative farmers of Ekdala seem to have been influenced by the success of new technology in neighbouring villages.

The dynamics of adoption has distributional implications. There is evidence indicating increasing concentration of control of various elements of the new technology (areas under irrigation and HYVs). One further problem is that farmers report the need for increasing fertilizer requirements to maintain yield. If this is the trend, then smaller farms' shortage of working capital may find it more difficult to maintain yield *vis-à-vis* that of larger farms. This may further accentuate the inequality in the distribution of incremental production. Thus the evidence of eventual equalization of adoption rates of HYV technology by various farm categories may not in itself be an insurance against increased inequality in the distribution of income and ownership of assets. This seems to be consistent with the view of Feder *et al.* (1985, p.288). It is also consistent with the view of Hayami and Ruttan (1984, p.49). In a community already characterized by an inegalitarian distribution of resources, the introduction of new technology is likely to reinforce existing inequality.

At any point of time (see Alauddin and Tisdell, 1988e) as, for instance, found by Muthia (1971) Schluter (1971) and Sharma (1973) in India and by I. Ahmed (1981) and Asaduzzaman (1979) in Bangladesh, the intensity of adoption of HYVs and related technology tends to be higher for small farms than for large ones, for those farms adopting HYV. However, the present study does not find evidence in Bangladesh to support Schluter's contention (1971) for India that this relationship becomes more marked with the amount of time that elapses following the introduction of the new varieties. Indeed, an opposite trend is apparent in Ekdala and in South Rampur where the indices of concentration of control of the areas under HYVs and irrigation seem relatively stable. It is possible that after the adoption of HYVs, concentration ratios tend at first to increase and eventually come to relatively stationary (equilibrium) values.

Recent studies have examined the relationship between agrarian innovations and rural poverty in LDCs, concentrating on either exchange (Ahluwalia, 1985; Mellor, 1985) or non-exchange income (Jodha, 1985; 1986). It is argued that neither approach employed independently is adequate. Evidence of growing concentration of control of land, of components of the new agricultural technology and of ancillary resources are documented, using Bangladeshi agricultural census and farm-level data. Access to land and other natural resources by the rural poor is gradually diminishing. Increasing landlessness and near-landlessness has resulted in greater dependence on wage employment for subsistence.

However, agricultural wages, being close to subsistence level, provide little scope for carry-over into periods of reduced agricultural activity. Even though there is some evidence of a slightly increasing trend in real wages, much of its effect on rural poverty is neutralized because of seasonality in employment and real wages. It is emphasized that the non-exchange component of income is important in off seasons and may become critical in abnormal years when both real wage and employment fall sharply (A.R. Khan, 1984). With rapid population growth and depletion of natural resources, and greater penetration of technological and market forces, the cushioning effect on the rural poor of access to natural resources in adverse economic circumstances has become more limited. Their income security has been undermined. These aspects of rural poverty have not been given much attention in previous studies of Bangladeshi agriculture.

Recent studies (e.g. Hazell, 1982; 1985; Mehra, 1981) suggest that the Green Revolution has been a source of increased variability of agricultural production. However, examination of Bangladeshi data employing an alternative methodology based on moving averages of production and yield suggests that the Green Revolution may have resulted in a reduction of relative variability of foodgrain production.

Even if variability should be higher during the original cropping period, the extra cropping in each year in most cases is likely to increase the degree of stability in annual production and yield. Even though absolute variability might in some cases show a tendency to increase, increased (annual) average production and yield can bring about a decrease in relative variability.

Examination of Bangladeshi national time series and district-level cross-sectional data in Chapter 10, employing Hazell's method, indicates that the Green Revolution may (in contrast to Hazell's findings for India) have reduced relative variability of foodgrain production and yield. Districts with higher adoption rates of HYVs and associated techniques seem to have lower relative variability. Furthermore, the probability of production and yield falling a certain percentage below the trend seem to be lower for high HYV adoption districts than those with lower adoption rates. Both increased incidence of multiple cropping and the greater use of modern agricultural techniques, such as irrigation, involving more control over production, are closely associated with declines in relative variability of crop production and yields. Increased use of controlled complementary inputs like chemical fertilizers and irrigation might also have reinforced the moderating impact of multiple cropping. Because these factors have been significant in Bangladesh, the Green Revolution appears to have had a stabilizing influence on the relative variability of production, rather than a destabilizing one.

For Bangladesh as a whole, an inter-temporal analysis indicates falling relative variability of foodgrain production and yield, and this is also true for most districts in Bangladesh. The results (both statistical and circumstantial) reported here for Bangladesh provide some evidence for the hypothesis that the Green Revolution has had a moderating impact on the relative foodgrain production and yield variability. At least this has been so for Bangladesh where cereal production is dominated by rice.

Chapter 11 extended research reported in Chapter 9 and 10. Research was extended by considering the apparent impact of the introduction of high-yielding varieties (HYVs) in Bangladesh on the lower semi-variance of foodgrain yields, on its lower semi-variation, and the probability of disaster yield levels, using a modification of Cherbychev's inequality. The trend in these characteristics is downwards and this is highlighted by a favourable shift in the empirically derived mean yield/risk efficiency locus. A number of factors may explain this shift. Using portfolio diversification theory and drawing on survey evidence, particular attention is given to greater crop diversification as a *possible* explanation, but it is suggested that it is not important in practice. Indeed, in the long term, reduced genetic diversity seems likely and may make it increasingly difficult to sustain agricultural yields (Oldfield, 1989).

Bangladesh's food supply problem will remain critical into the foreseeable future, because the overall rate of its population growth is outstripping growth rate of its food production. While our *revised series* of foodgrain production gives a more optimistic picture of foodgrain growth rates (2.87 per cent) for the 1965–6 to 1980–1 period, there are doubts about the reliability of the *revised series*, based on the hypothesis of Boyce (1985). Agricultural production as a whole (that is, including foodgrains and other crops) is failing to keep pace with population growth, whether the growth rate of agricultural production estimated from official data for 1965–80 is used (1.52 per cent) or Boyce's revised estimate of 2.18 per cent is adopted.

The crucial question is whether or not the recent growth rates can be sustained. The restoration of Bangladesh's food supply–demand balance depends critically on a significant reduction in population growth and an increase in yield per hectare. While the agronomic potential for increasing yield is substantial, its realization requires increased intensity of cropping. But there may be ecological limits to the degree of intensification that is sustainable. Also, account needs to be taken of major socio-economic constraints on achievement of agronomic potential.

In a wider context, Bangladesh, along with other countries, may encounter problems in the long run in sustaining agricultural productivity because of loss of genetic diversity, ecological damage due to application of agro-chemicals, and from its dependence on chemical products (e.g. fertilizers and pesticides) produced from exhaustible and non-renewable resources. Another difficulty for Bangladesh is that increased food production using modern techniques tends to raise its dependence on imported technology and imported inputs. For Bangladesh, a great number of modern agricultural inputs are imported and even those that are domestically produced are highly import intensive.

It is unclear whether the new agricultural technology introduced in Bangladesh will result in a sustainable higher production per capita than in the past. The present study has attempted to identify obstacles to such sustainability. However, it is conceivable that there may be major breakthroughs in new agricultural technology in the future. Nevertheless, continuing concern about the issues involved is not misplaced (Boyce, 1987a).

The conclusions of this study, while substantial and revealing, need to be judged in the light of the following caveats: First, the imperfect quality of the secondary data which form the statistical basis of most chapters of this study must be borne in mind, particularly in the light of discussion in Chaptere 12. Secondly, the primary data used in chapter 7, Chapter 9 and Chapter 13 are based on a limited number of observation from only two villages. Its limited scope, micro-nature and regional orientation make generalizations difficult.

However, the findings are indicative of the forces at work in the process of agricultural development in Bangladesh following the Green Revolution. Further studies paying particular attention to the issues mentioned above, employing broad-based farm-level and secondary data involving aspects of agricultural research and development, would be useful in strengthening the present knowledge about impacts of the 'Green Revolution' discussed in the book.

Notes

Chapter 1

1. For an excellent survey of industrialization and employment in LDCs see Morawetz (1974).
2. See Shand (1985) for the importance of off-farm employment in the development of rural Asia.
3. Closely related to this strategy was the priority or lack of it accorded to agriculture in the development plans. Even though every plan claimed to have given top priority to agriculture, statistical evidence does not seem to substantiate this. For details see Huq (1963, pp.157–8) and Papanek (1967, pp.165–6).
4. For a brief history of the process of institutionalization of agricultural research, its organization and other relevant aspects see Alauddin (1981); Pray and Anderson (1985).

Chapter 2

1. Early 1969 (i.e., later part of 1968–9) marks the end of the 'Ayub regime' which was succeeded in March 1969 by another military take-over.
2. For a generalized form of kinked exponential model see Boyce 1986, p.387; see also Boyce 1987b).
3. The way in which Venegas and Ruttan present their method is slightly different from those in the present study (Venegas and Ruttan, 1964, p.161). They present area and yield effects as follows:

$$(\Delta \log A / \Delta \log P) \times 100 + (\Delta \log Y / \Delta \log P) \times 100 = 100 \text{ per cent} \quad \text{(2A-1)}$$
$$\text{(Area effect)} \qquad\qquad \text{(Yield effect)}$$

where, $\Delta \log A = \log A_t - \log A_0$, etc. These two effects add up to 100 per cent, except for some very small balancing term, i.e. logarithmic interpolation term. Notwithstanding the apparent dissimilarity, (2A-1) is mathematically equivalent to expression (2.2) used above.
4. See Little (1960, pp.217–37) for a comprehensive survey of this debate.
5. In the view of the present authors it would be desirable to divide extra area used for a crop into that obtained by an increase in the frequency of cropping and that stemming from employment of land not previously used for cultivation of crops. It should be possible to do this in principle, even though more data would be required than for the decomposition methods discussed.
6. Such breakdowns are not available for individual years of the 1960s.
7. The original figures for agricultural value added for 1967–8, 1968–9 and 1969–70 at constant 1959–60 factor-cost were available from Alamgir and Berlage (1974, p.168). EIU (1976, p.17) estimates of 1969–70 agricultural

value added are at 1972–3 prices. The late 1960s figures were converted into 1972–3 prices using the appropriate sectoral GDP deflator. The crop sub-sector's contribution to agricultural value added ranged from 75 to 78 per cent during that period. A factor of 76 per cent has been used to derive value added from crops. The 1982–4 value-added figure was derived from information contained in BBS (1986d, p.167). The per hectare value-added figures were estimated as ratios of the total figures for value added from crops to those of net cropped area for the respective periods.
8. As noted in the last section, a higher proportion of increased output is assigned to yield with greater aggregation (e.g. for rice). This is indicative of the fact that *annual* yields have risen, even though seasonal yields (e.g. for *aus* HYV) have fallen. Therefore the aggregate data may give a more realistic picture of the sources of growth.

Chapter 3

1. de Janvry (1978, p.307) distinguishes four types of agricultural innovations: (1) biological; (2) mechanical; (3) chemical and (4) agronomic.
2. For example, the nutrient contents of Urea, TSP and MP are 46.0, 46.0 and 60.0 per cent of the respective absolute quantities (BBS, 1984b, p.227).

Chapter 4

1. The time cut-off points employed here are different from those in Chapter 2 for two reasons. First, even though national data on all HYVs are available since 1967–8, district-level data on all HYVs are available only from 1969–70. Secondly, at a more disaggregated level, it is difficult to justify 1975–6 as a uniform dividing line. This is because the impact of natural and other factors is unlikely to be as clear-cut for each district as it is likely to be for Bangladesh as whole. Hence no attempt is made to divide the period of the new technology into two sub-periods as was done in Chapter 2.
2. Modern methods include irrigation by shallow and deep tubewells (STW, DTW), low lift pumps (LLP) and large-scale canals.
3. Consider the correlation coefficients involving the three explanatory variables: FERT and PRHYVF = 0.6706; PRFAI and FERT = 0.7038; PRHYVF and PRFAI = 0.5665.
4. See, *inter alia*, Bardhan (1973); Bharadwaj (1974); Berry and Cline (1979); Chattapadhyay and Rudra (1976); Cheung (1968); Ghose (1979); M. Hossain (1977, 1978); Jabbar (1977); Jannuzi and Peach (1980); and A.K. Sen (1966, 1975).
5. On cash-cost basis, owner cultivators' net returns per hectare for HYVs of wheat, *boro*, transplanted and broadcast *aus* and transplanted *aman* crops of HYV rice are 533, 1391, 1951, 1924 and 1947 taka respectively (Alauddin and Mujeri, 1986a, p.71).

Chapter 7

1. For an excellent survey of the literature on adoption and diffusion of agricultural innovations see Feder *et al.* (1985).
2. Kislev and Schchori-Bachrach (1973) emphasize the role of skill and entrepreneurship in the diffusion process.
3. These are not the only possible models. Kautsky (1899) (see also Banaji, 1976), for example, sees innovation in agriculture as a flow-on from the penetration of industry to country towns. Farmers' cash needs increase so as to purchase farm capital produced by such industry, and other dynamic changes occur which make farmers more dependent on the market. In this way they may be subjected to increased risks, and the uncertainty elements in the decision-making become more important. Neoclassical economists such as Marshall (1890) and Hicks (1946) give little attention to uncertainty in their decision-making models, unlike the safety-first models.
4. This method has been widely used in the biological sciences. See, for example, Finney (1971).
5. For a more advanced treatment see Kendall (1980, pp.145–69).
6. However, *Paijam* is not officially recognized as an HYV (Pray and Anderson, 1985, p.43). Clay (1986, p.179) notes that *Paijam* is one of the most successful modern rice varieties, while Pray and Anderson (1985, p.45) indicate that it spread rapidly, despite recommendations of agricultural extension agent that it should not be used.
7. Similar evidence of replacement of earlier varieties by new ones has also been found in other areas of Bangladesh (see, for example, Jones, 1984, pp.198–9). See also Jain *et al.* (1986, p.84).
8. Cf. Rogers and Shoemaker (1971, pp.184–5).
9. This is somewhat similar to the dependence of farmers on informal sources of R & D. For details see Biggs and Clay (1981).

Chapter 8

1. For instance, 'there was a strict objective definition of farm household in 1983–84. It is possible that due to subjective procedure followed in 1977 many of the small farm households were not considered as farm holdings' (BBS, 1986b, p.32–3).

Chapter 9

1. Some earlier studies (S.R. Sen, 1967; Rao, 1975) indicated causal connection between agricultural development and instability in agricultural output growth in India. For evidence of increased yield variability in cereal grains see Barker *et al.* (1981).
2. See also Rudra (1970) for further discussion on various functional forms for estimating growth rates.
3. Dantwala (1985, pp.112, 123) cites an official report which seems to substantiate the Sarma and Roy (1979) findings. Thus: 'it is noteworthy that fluctuations in paddy production have been higher in areas where

new technology has not yet become fully entrenched. This is evident from the analysis of regionwise decline in output during the drought years 1972–73, 1976–77 and 1979–80. During these three years, decline in paddy production due to below normal monsoon was the lowest in Haryana and Punjab. The most seriously affected States were the traditional paddy growing areas where the new technology is yet to make a significant mark. In the case of wheat, the production in Punjab and Haryana in fact increased during these years of drought' (Government of India, 1982, quoted in Dantwala, 1985, p.123).
4. The average production of rice in South Asia as defined by Hazell (1985) during 1971–2 to 1982–3 was 23 347 thousand tonnes (Hazell, 1985, p.150). Available sources (mentioned in the text) suggest the corresponding figure for Bangladesh to be 12 463 thousand tonnes.
5. The average share of rice in total foodgrains was more than 96 per cent during 1971–2 to 1982– 3.
6. Similar evidence is provided by Murshid (1986) who examined variability of Bangladeshi foodgrain production for the 20-year period 1960–1 to 1979–80. There did not appear to be any trend in instability even though it showed some tendency to increase in the early 1970s (p.66).

Chapter 10

1. Semi-log trend was also fitted but the results were not any different in terms of R^2 and t-values.
2. The definition of coefficient of variation employed in Tables 10.1, 10.2 and 10.3 does not conform to the standard definition used in statistical literature. As deviations are measured around the trend values instead of the actual mean values, it is perhaps more appropriate to term it coefficient of 'unexplained' variation rather than coefficient of variation.
3. This seems to be generally true of relative production and yield variability of HYVs. Coefficients of variation of production and yield of *kharif* and *rabi* HYVs considered separately for districts in Bangladesh indicated greater relative production and yield variability compared with all HYVs taken together. That is, on an annual basis HYVs show lower relative variability than on a seasonal basis.
4. It might also be interesting to see whether irrigation (PRFAI) and fertilizer application (FRTHEC) have had any moderating impact upon relative production and yield variability (CVPROD and CVYIELD) in the period of the new technology. The following equations do not indicate any strong negative relationships, even though a weakly moderating impact appears to show up:

$$CVPROD = 10.8721 - 0.0192PRFAI$$
$$(R^2 = 0.0016,\ t = 0.150) \tag{10A.1}$$

$$CVPROD = 12.9560 - 0.0937FRTHEC$$
$$(R^2 = 0.1935,\ t = 1.900) \tag{10A.2}$$

$$\text{CVYIELD} = 8.9302 - 0.1102\text{PRFAI}$$
$$(R^2 = 0.1154, t = 1.400) \tag{10A.3}$$

$$\text{CVYIELD} = 10.3260 - 0.0709\text{FRTHEC}$$
$$(R^2 = 0.2459, t = 2.310) \tag{10A.4}$$

5. Consider the following:

$$\text{PRHYVF} = 13.3980 + 0.3155\text{POTKHYV}$$
$$(R^2 = 0.2118, t = 2.010) \tag{10A.5}$$

$$\text{PCKHARIF} = -3.0452 + 0.3920\text{POTKHYV}$$
$$(R^2 = 0.3257, t = 2.690) \tag{10A.6}$$

where PRHYVF and POTKHYV have the same meanings as before but relate to the average for the years 1982–3, 1983–4 and 1984–5; PCRKHARIF refers to the average percentage of *kharif* foodgrain areas in HYV for the 1982–3 to 1984–5 period.

6. It is interesting to see if production and yield variability during the earlier period (CVP1 and CVY1) have a relationship with those of the subsequent years (CVP2 and CVY2) or with the extent of adoption of HYVs (PRHYVF). To examine this, Equations (10A.5) to (10A.8) were estimated. Equations (10A.5) and (10A.6) which show the relationship between variabilities during the pre- and post-Green Revolution periods do not indicate any strong impact of the variability of the earlier period on those of the latter. Equations (10A.7) and (10A.8) estimate the relationship between the production and yield variability of the pre-Green Revolution period and the per cent area under HYV foodgrains. They do not show any statistically significant relationship between the two sets of variables. Thus Equations (10A.5) to (10A.8) do not provide strong support for the hypothesis that a positive association exists between the variability of the earlier period and that of the latter, or with the extent of HYV adoption. But *a fortiori* there is no evidence to suggest that they are negatively related.

$$\text{CVP2} = 7.0558 + 0.2186\text{CVP1}$$
$$(R^2 = 0.0682, t = 1.050) \tag{10A.7}$$

$$\text{CVY2} = 2.5448 + 0.4056\text{CVY1}$$
$$(R^2 = 0.2337, t = 2.140) \tag{10A.8}$$

$$\text{PRHYVF} = 28.1632 - 0.5285\text{CVP1}$$
$$(R^2 = 0.0293, t = 0.670) \tag{10A.9}$$

$$\text{PRHYVF} = 25.2952 - 0.4667\text{CVY1}$$

$$(R^2 = 0.0102, t = 0.390) \tag{10A.10}$$

7. During 1969–70 to 1984–5, rice constituted nearly 97 per cent of total foodgrain production in Bangladesh. In India on the other hand the corresponding figure for rice during 1967–8 to 1977–8 was 44.8 per cent (Hazell, 1982, p.13). Other important food crops in India during the same period were wheat (25.6 per cent), *jowar* (10 per cent), maize (6.4 per cent) and *bajra* (5.4 per cent).

Chapter 12

1. Boyce (1985) fitted an exponential trend. Boyce (1986, p.389), using a kinked exponential model, estimates annual agricultural output growth rates of 1.57 and 2.49 per cent respectively for the 1949–64 and 1965–80 periods.
2. Net availability of foodgrains for consumption is defined as gross production less 10 per cent for seed, feed and wastage, less internal procurement plus off-take from government ration distribution. Off-take from government ration distribution is made up of imports and internal procurement and carry-overs from the previous year.

Chapter 13

1. For details see Douglass (1984) and Tisdell (1988).
2. Net cultivated area (NCA), that is the area cultivated only once during the year, increased from about 8.0 million ha in the late 1940s to about 8.4 million ha on average throughout the 1950s, and by the early 1960s it reached a little over 8.5 million ha (EPBS, 1969, p.41). As can be seen from Table 13.1, the upward trend in net cultivated area continued until the end of the 1960s when it reached a peak of 8.8 million ha. Since the 1970s, net cropped area has shown clear signs of decline. Even though net cultivated area increased marginally to 8.6 million ha in the early 1980s, it was still lower than that in the late 1960s, so that in the intervening period some land was 'lost' from cultivation. Gross annual cultivated area (GCA), that is, net cultivated area plus area cultivated more than once during the year, remained stagnant at around 13.5 million ha in the 1950s and rose to about 13.0 million ha by the late 1960s. By the earlier part of 1980s it increased to a little over 13.2 million hectares. Both net and gross cultivated land per head of population (NCAPHD and GCAPHD) have declined monotonically over the years, the fall in the former being more rapid than that for the latter. The trends in gross and net cultivated land in aggregate, as well as per capita terms, can be specified by the regressions given in Equations 13.4 to 13.7. Overall, net cropped area and gross cropped area have expanded, but net cropped area and gross cropped area per head have declined at a decreasing rate with the passage of time.

$$NCA = 8.3655 + 0.0062T$$
$$(R^2 = 0.1348, t\text{-value} = 2.270) \tag{13A.1}$$

$$GCA = 10.4519 + 0.0798T$$
$$(R^2 = 0.7296, t\text{-value} = 9.440) \tag{13A.2}$$

$$1n\text{NCAPHD} = -1.6510 - 0.0234T$$
$$(R^2 = 0.9909, t\text{-value} = 60.07) \tag{13A.3}$$

$$1n\text{GCAPHD} = -1.4240 - 0.0173T$$
$$(R^2 = 0.9385, t\text{-value} = 22.44) \tag{13A.4}$$

3. This was confirmed in an interview with a soil scientist at the Bangladesh Rice Research Institute (BRRI). Jones (1984) also reports zinc and sulphur deficiency caused by multiple cropping in the Dhaka district of Bangladesh.
4. This is consistent with the findings of R.I. Rahman (1983) who cites farm-level evidence from Dhaka district of Bangladesh. As Rahman puts it (p.65), the farmers, 'need to use more fertilizer to achieve the same increase in yield or even to maintain the same yield. . .a higher dose of fertilizer to maintain the same yield may simply mean that loss of yield due to deterioration of the quality of seed or loss of land fertility due to higher cropping intensity is compensated by fertilizer use'.
5. One needs also to consider the substantial costs involved in heavy dependence on imported technology. Apart from requirements of foreign exchange, technologies (fertilizers, pesticides and insecticides and irrigation machinery) materially transferred to countries with characteristics different from those in the country of origin are unlikely to perform in the way they have performed in the area to which they are indigenous (Bruton, 1955, p.336). Even though agro-chemicals may be applicable to a wide range of environmental conditions, one needs to consider a number of factors more or less unique to agricultural innovations. Differences in ecological conditions influenced by geophysical and climatic variables have a critical influence on the performance of agricultural innovations in farmers' fields. This highlights the importance in agriculture of diversity or heterogeneity in environmental conditions (for further details, see Alauddin and Tisdell, 1987; Biggs and Clay, 1981).
6. Intermediate input level: using a basic package of fertilizers and biocides with some improved crop varieties, simple long-term conservation measures and and existing crop mixes on half the land and the most productive crop mix. A high input level involves full use of fertilizer and biocides, improved crop varieties, conservation measures and the best mix of crops on all the land, and is approximately equivalent to the Western European level of input use (FAO, 1984, pp.2–3, 15).

Bibliography

Note: For each author, collaborated works are listed alphabetically by second author, regardless of the number of co-authors.

ABDULLAH, A.A. (1976) 'Land reform and agrarian change in Bangladesh', *Bangladesh Development Studies*, 4(1), pp.67–114.

ABDULLAH, A.A., HOSSAIN, M. and NATIONS, R. (1976) 'Agrarian structure and IRDP: Preliminary considerations', *Bangladesh Development Studies*, 4(2), pp.209–66.

ACO (1962) *Pakistan Census of Agriculture 1960: Final Report – East Pakistan*, vol. I, part I, Karachi: Agricultural Census Organization, Ministry of Food and Agriculture, Government of Pakistan.

AHLUWALIA, M.S. (1978) 'Rural poverty and agricultural peformance in India', *Journal of Development Studies*, 14(3), pp.298–323.

—— (1985) 'Rural poverty, agricultural production and prices: A reexamination', in Mellor and Desai (eds) (1985), pp.59–75.

AHMAD, K. and HASSAN, N. (eds) (1983) *Nutrition Survey of Rural Bangladesh, 1981–82*, Dhaka: Institute of Nutrition and Food Science, University of Dhaka.

AHMED, I. (1981) *Technological Change and Agrarian Structure: A Study of Bangladesh*, Geneva: International Labour Office.

AHMED, M. (1986) 'The use and abuse of pesticides and the protection of environment', in BMOE (1986), pp.89–95.

AHMED, M.F. (1986) 'Modern agriculture and its impact on environmental degradation', BMOE (1986), pp.42–9.

AHMED, R. (1978) 'Price support versus input subsidy for increasing rice production in Bangladesh', *Bangladesh Development Studies*, 6(2), pp.119–38.

—— (1979) *Foodgrain Supply, Distribution, and Consumption Policies within a Dual Pricing Mechanism: A Case Study of Bangladesh*, Research Report no. 8, Washington, D.C.: International Food Policy Research Institute.

—— (1981) *Agricultural Price Policies under Complex Socio-Economic and Natural Constraints: The Case of Bangladesh*, Research Report no. 27, Washington, D.C.: International Food Policy Research Institute.

ALAM, M. (1977) 'A report on BADC-owned deep tubewell irrigation in Dhaka and Khulna divisions', Dhaka: Bangladesh Institute of Development Studies (mimeo).

—— (1984) *Capital Accumulation and Agrarian Structure in Bangladesh: A Study on Tubewell Irrigated Villages of Rajshahi and Comilla*, Dhaka: Dhaka University Centre for Social Studies.

ALAMGIR, M. (1980) *Famine in South Asia*, Cambridge, Mass.: Oelgeschlager, Gunn & Hain.

ALAMGIR, M. and BERLAGE, L.J.J.B. (1973) 'Estimation of income elasticity of demand for foodgrain in Bangladesh from cross section data:

A skeptical view', *Bangladesh Economic Review*, 1(4) pp.25–58.

—— (1974) *Bangladesh National Income and Expenditure, 1949–50 to 1969–70*, Dhaka: Bangladesh Institute of Development Studies.

ALAUDDIN, M. (1981) 'Agricultural research organization and policy in Bangladesh', *Bangladesh Journal of Agricultural Economics*, 4(2), pp.1–24.

—— (1988) *The 'Green Revolution' in Bangladesh: Implications for Growth, Distribution, Stability and Sustainability*, unpublished PhD thesis, Newcastle, Australia: University of Newcastle.

ALAUDDIN, M. and MUJERI, M.K. (1981) 'The strategy of agricultural development in Bangladesh: Past and present', *Bangladesh Journal of Political Economy*, 5, pp.235–53.

—— (1985) 'Employment and productivity growth in Bangladesh agriculture: An inter-district analysis', *Marga*, 8(1), pp.50–72.

—— (1986a) 'Growth and change in the crop sector of Bangladesh: A disaggregated analysis', *Journal of Contemporary Asia*, 16(1), pp.54–73.

—— (1986b) 'Growth and changes in the crop sector of Bangladesh: A review of performance and implications for policies', *Scandinavian Journal of Development Alternatives*, 5(4), pp.109–25.

ALAUDDIN, M. and TISDELL, C.A. (1986a) 'Bangladeshi and international agricultural research: Administrative and economic issues', *Agricultural Administration*, 21(1), pp.1–20.

—— (1986b) 'Inappropriate industries and inefficient resource-use in Bangladesh: Some evidence from input–output analysis', *Socio-Economic Planning Sciences*, 20(3), pp.135–43.

—— (1986c) 'Decomposition methods, agricultural productivity growth and technological change: A critique supported by Bangladeshi data', *Oxford Bulletin of Economics and Statistics*, 48(4), pp.353–72.

—— (1986d) 'Market analysis, technical change and income distribution in semi-subsistence agriculture: The case of Bangladesh', *Agricultural Economics*, 1(1), pp.1–18.

—— (1987) 'Trends and projections of Bangladeshi food production: An alternative viewpoint', *Food Policy*, 12(4), pp.318–31.

—— (1988a) 'The use of input–output analysis to determine the appropriateness of technology and industries: Evidence from Bangladesh', *Economic Development and Cultural Change*, 36(2), pp.369–92.

—— (1988b) 'New agricultural technology and sustainable food production: Bangladesh's achievements, predicaments and prospects', in C.A. Tisdell and P. Maitra (eds) (1988), pp.34–61.

—— (1988c) 'Has the 'Green Revolution' destabilized foodgrain production? Some Evidence from Bangladesh', *Developing Economies*, 26(2), pp.141–60.

—— (1988d) 'Impact of new agricultural technology on the instability of foodgrain production and yield: Data analysis for Bangladesh and its districts', *Journal of Development Economics*, 29(2), pp.199–227.

—— (1988e) 'Patterns and determinants of adoption of high yielding varieties: Farm-level evidence from Bangladesh', *Pakistan Development Review*, 27(2), pp.183–227.

—— (1988f) 'Dynamics of adoption and diffusion of HYV technology: New evidence of inter-farm differences in Bangladesh', *Research Report or Occasional Paper No. 155*, University of Newcastle, Australia: Department of Economics.

—— (1988g) 'Bangladeshi crop production and food supply: Growth rates and changing composition in response to new technology', *Research Paper No. 210*, Department of Economics, University of Melbourne.

—— (1989a) 'Bio-chemical technology and Bangladeshi land productivity: Diwan and Kallianpur's analysis reapplied and critically examined', *Applied Economics*, 21(6), pp.741–60.

—— (1989b) 'Poverty, resource distribution and security: The impact of new agricultural technology in rural Bangladesh', *Journal of Development Studies*, 25(4), pp.550–70.

—— (1989c) 'The 'Green Revolution' and labour absorption in Bangladesh agriculture: Relevance of East Asian experience', *Research Report No. 229*, University of Melbourne: Department of Economics.

—— (1989d) 'Rural poverty and resource distribution in Bangladesh: "Green Revolution" and beyond', in B. Greenshields and M. Bellamy (eds.), pp.247–53.

ALLEN, R. (1980) *How to Save the World*, London: Kogan Page.

ANDERSON, J.R., DILLON, J.L. and *HARDAKER, J.B.* (1977) *Agricultural Decision Analysis*, Ames, Iowa: Iowa State University Press.

ANTLE, J.M. (1983) 'Infrastructure and aggregate agricultural productivity', *Economic Development and Cultural Change*, 31(4), pp.609–19.

ASADUZZAMAN, M. (1979) 'Adoption of HYV rice in Bangladesh', *Bangladesh Development Studies*, 7(3), pp.23–52.

BADC (1978) *Annual Report, 1977–78*, Dhaka: Bangladesh Agricultural Development Corporation.

—— (1979) *Annual Report, 1978–79*, Dhaka: Bangladesh Agricultural Development Corporation.

—— (1980) *Annual Report, 1979–80*, Dhaka: Bangladesh Agricultural Development Corporation.

—— (1981) *Annual Report, 1980–81*, Dhaka: Bangladesh Agricultural Development Corporation.

—— (1984) *Annual Report, 1982–83*, Dhaka: Bangladesh Agricultural Development Corporation.

—— (1985) *Annual Report, 1983–84*, Dhaka: Bangladesh Agricultural Development Corporation.

BALYSS-SMITH, T.P. AND WANMALI, S. (eds) (1984) *Understanding Green Revolutions: Agrarian Change and Development Planning in South Asia*, Cambridge: Cambridge University Press.

BANAJI, J. (1976) 'Summary of selected parts of Kautsky's *The Agrarian Question*', *Economics and Society*, 5(1), pp.2–49.

BARBIER, E.B. (1987) 'The concept of sustainable economic development', *Environmental Conservation*, 14(2), pp.101–10.

BARDHAN, P.K. (1973) 'Size, productivity and returns to scale: An analysis of farm level data in Indian agriculture', *Journal of Political Economy*, 81(6), pp.1370–86.

—— (1984) *Land, Labor and Rural Poverty: Essays in Development Economics*, Delhi: Oxford University Press.

BARKER, R., GABLER, E.C. and WINKELMANN, D. (1981) 'Long-term consequences of technological change on crop yield stability: The case for cereal grain', in A. Valdes (ed.), *Food Security for Developing Countries*, Boulder, Colorado: Westview Press, pp.53–78.

BARKER, R. and HERDT, R.W. (1978) 'Equity implications of technological change', in IRRI (1978), pp.83–108.

BBS (1972) *Master Survey of Agriculture in Bangladesh* (7th Round, 2nd Phase), Dhaka: Bangladesh Bureau of Statistics.

—— (1976) *Agricultural Production Levels of Bangladesh, 1947–72*, Dhaka: Bangladesh Bureau of Statistics.

—— (1978) *Statistical Pocket Book of Bangladesh, 1978*, Dhaka: Bangladesh Bureau of Statistics.

—— (1979) *Statistical Year Book of Bangladesh, 1979*, Dhaka: Bangladesh Bureau of Statistics.

—— (1980a) *Year Book of Agricultural Statistics of Bangladesh, 1979–80*, Dhaka: Bangladesh Bureau of Statistics.

—— (1980b) *The Household Expenditure Survey of Bangladesh, 1973–74*, Dhaka: Bangladesh Bureau of Statistics.

—— (1981) *Report on the Agricultural Census of Bangladesh, 1977 (National Volume)*, Dhaka: Bangladesh Bureau of Statistics.

—— (1982) *Statistical Year Book of Bangladesh, 1982*, Dhaka: Bangladesh Bureau of Statistics.

—— (1984a) *Monthly Statistical Bulletin of Bangladesh*, March, Dhaka: Bangladesh Bureau of Statistics.

—— (1984b) *Statistical Year Book of Bangladesh, 1983–84*, Dhaka: Bangladesh Bureau of Statistics.

—— (1985a) *1983–84 Year Book of Agricultural Statistics of Bangladesh*, Dhaka: Bangladesh Bureau of Statistics.

—— (1985b) *1984–85 Statistical Year Book of Bangladesh*, Dhaka: Bangladesh Bureau of Statistics.

—— (1985c) *1984–85 Year Book of Agricultural Statistics of Bangladesh*, Dhaka: Bangladesh Bureau of Statistics.

—— (1986a) *Monthly Statistical Bulletin of Bangladesh*, May, Dhaka: Bangladesh Bureau of Statistics.

—— (1986b) *The Bangladesh Census of Agriculture and Livestock: 1983–84: Volume I, Structure of Agricultural Holdings and Livestock Population*, Dhaka: Bangladesh Bureau of Statistics.

—— (1986c) *The Bangladesh Census of Agriculture and Livestock: 1983–84: Volume II, Cropping Patterns*, Dhaka: Bangladesh Bureau of Statistics.

—— (1986d) *Monthly Statistical Bulletin of Bangladesh*, November, Dhaka: Bangladesh Bureau of Statistics.

—— (1986e) *Bangladesh Household Expenditure Survey, 1981–82*, Dhaka: Bangladesh Bureau of Statistics.

BERA, A.K. and KELLEY, T.G. (1990) 'Adoption of high yielding rice varieties in Bangladesh: An econometric analysis', *Journal of Development Economics*, 33, pp.263–85.

BERRY, R.A. and CLINE, W.R. (1979) *Agrarian Structure and Productivity in Developing Countries*, Baltimore, Md.: Johns Hopkins University Press.

BHARADWAJ, K. (1974) *Production Conditions in Indian Agriculture: A Study Based on Farm Management Surveys*, London: Cambridge University Press.
BIGGS, S.D. (1982) *Agricultural Research: A Review of Social Science Analysis*, Discussion Paper no. 115, Norwich: University of East Anglia School of Development Studies.
BIGGS, S.D. and CLAY, E.J. (1981) 'Sources of innovation in agricultural technology', *World Development*, 9(4), pp.321–36.
BINSWANGER, H.P. and RUTTAN, V.W. (eds) (1978) *Induced Innovation: Technology, Institutions and Policy*, Baltimore, Md.: Johns Hopkins University Press.
BISHOP, R. (1978) 'Endangered species uncertainty: The economics of a safe minimum standard', *American Journal of Agricultural Economics*, 60(1), pp.10–18.
BMAF (1979) 'Costs and returns survey for Bangladesh 1978–79 crops: vol. II, Broadcast *aman* paddy', Dhaka: Bangladesh Ministry of Agriculture and Forests, Agro-Economic Research Section.
—— (1981a) '1980–81 T. *aman* Cultivation in Bangladesh: An economic profile', Dhaka: Bangladesh Ministry of Agriculture and Forests, Agro-Economic Research Section.
—— (1981b) 'Costs and returns survey for Bangladesh 1980–81 crops: vol. II, *Boro* paddy', Dhaka: Agro-Economic Research Section, Bangladesh Ministry of Agriculture and Forests.
BMOE (1986) *Protection of Environment from Degradation*, Proceedings of South Asian Association for Regional Cooperation (SAARC) Seminar 1985, Dhaka: Bangladesh Ministry of Education, Science and Technology Division.
BOYCE, J.K. (1985) 'Agricultural growth in Bangladesh, 1949–50 to 1980–81: A review of evidence', *Economic and Political Weekly*, 20(13), pp.A31–43.
—— (1986) 'Kinked exponential models for growth rate estimation', *Oxford Bulletin of Economics and Statistics*, 48(4), pp.385–91.
—— (1987a) 'Trends and projections for Bangladeshi food production: Rejoinder to M. Alauddin and C. Tisdell', *Food Policy*, 12(4), pp.332–6.
—— (1987b) *Agrarian Impasse in Bengal: Institutional Constraints to Technological Change*, Oxford: Oxford University Press.
BPC (1980) *The Structure of the Bangladesh Economy: An Input–Output Analysis*, Background papers of the Second Five Year Plan of Bangladesh, vol. I, Dhaka: Bangladesh Planning Commission.
—— (1985) *The Third Five Year Plan 1985–90*, Dhaka: Bangladesh Planning Commission.
BRAMER, H. (1974) 'The potential for rainfed HYV rice cultivation in Bangladesh', in BRRI, *Workshop on Massive Production of HYV Rice under Rainfed Conditions in Bangladesh*, Joydevpur: Bangladesh Rice Research Institute, pp.30–76.
BROWN, L. (1970) *Seeds of Change*, New York: Praeger.
BRRI (1977) *Workshop on Ten Years of Modern Rice and Wheat Cultivation in Bangladesh*, Joydevpur: Bangladesh Rice Research Institute.
BRUTON, H.J. (1955) 'Growth models and underdeveloped economies',

Journal of Political Economy, 63(3), pp.322–36.

CAIN, M. (1983) 'Landlessness in India and Bangladesh: A critical review of national data sources', *Economic Development and Cultural Change*, 32(1), pp.149–68.

CHAKRABARTI, S. (1982) 'In search of growth pattern for foodgrain production in India', *Economic and Political Weekly*, 17(52), pp.A122–26.

CHAMBERS, R. (1984) 'Beyond the Green Revolution: A selective essay', in T.P. Balyss-Smith and S. Wanmali (eds.) (1984), pp.362–79.

CHAMBERS, R., LONGHURST, R. and PACEY, A. (eds) (1981) *Seasonal Dimensions of Rural Poverty*, London: Frances Pinter.

CHATTAPADHYAY, M. and RUDRA, A. (1976) 'Size-productivity revisited', *Economic and Political Weekly*, 11(39), pp.A104–16.

CHAUDHURY, R.H. (1981a) 'The seasonality of prices and wages in Bangladesh', in Chambers *et al.* (eds) (1981), pp. 87–92.

—— (1981b) 'Population pressure and agricultural productivity in Bangladesh', *Bangladesh Development Studies*, 9(3), pp.67–88.

CHEUNG, S.N.S. (1968) 'Private property rights and share-cropping', *Journal of Political Economy*, 76(6), pp.1108–22.

CHISHOLM, A.H. (1988) 'Sustainable resource use and development: Uncertainty, irreversibility and rational choice', in Tisdell and Maitra (eds) (1988), pp.188–216.

CHISHOLM, A.H. and TYRES, R. (eds) (1982) *Food Security: Theory, Policy and Perspectives from Asia and the Pacific Rim*, Lexington, Mass.: Lexington Books.

CIRIACY-WANTRUP, S.V. (1968) *Resource Conservation: Economics and Politics*, Division of Agricultural Services, Berkeley and Los Angeles: University of California.

CIRIACY-WANTRUP, S.V. and BISHOP, R. (1975) '"Common Property" as a concept in natural resource policy', *Natural Resources Journal*, 15(4), pp.713–27.

CLARKE, W.C. (1971) *Place and People: An Ecology of a New Guinean Community*, Canberra: Australian National University Press.

CLAY, E.J. (1981) 'Seasonal patterns of agricultural employment in Bangladesh', in Chambers *et al.* (eds) (1981), pp.92–101.

—— (1986) 'Releasing the hidden hand: A review of E. Boyd Wennergren, Charles H. Antholt and Maurice D. Whitaker, *Agricultural Development in Bangladesh*, *Food Policy*, 11(2), pp.178–80.

CLAY, E.J. *et al.* (1978) '*Yield assessments of broadcast aman (deep-water rice) in selected areas of Bangladesh in 1977*', Joydevpur: Bangladesh Rice Research Institute.

CONWAY, G. (1985) 'Agricultural ecology and farming systems research', in J.V. Remenyi (ed.) *Agricultural Systems Research for Developing Countries*, ACIAR Proceedings no. 11, Canberra: Australian Centre for International Agricultural Research, pp.43–9.

—— (1986) *Agroecosystem Analysis for Research and Development*, Bangkok: Winrock International.

CORNES, R., MASON, C.F. and SANDLER, T. (1986) 'The commons and the optimal number of firms', *Quarterly Journal of Economics*, 101(3), pp.641–7.

CUMMINGS, J.T. (1974) 'The supply responsiveness of Bangalee rice and cash crop cultivators', *Bangladesh Development Studies*, 2(4), pp.857–66.

DALRYMPLE, D.G. (1976) *Development and Spread of High Yielding Varieties of Wheat and Rice in the Less Developed Nations*, Foreign Agricultural Report no. 95, Washington, D.C.: US Department of Agriculture.

—— (1977) 'Evaluating the impact of international research on wheat and rice production in the developing nations', in T.M. Arndt, D.G. Dalrymple and V.W. Ruttan (eds), *Resource Allocation and Productivity in National and International Agricultural Research*, Minneapolis: University of Minnesota Press, pp.171–208.

—— (1985) 'The development and adoption of high-yielding varieties of wheat and rice in developing countries', *American Journal of Agricultural Economics*, 67(5), pp.1067–73.

DANTWALA, M.L. (1985) 'Technology, growth and equity in agriculture', in Mellor and Desai (eds) (1985), pp.110–23.

DE JANVRY, A. (1978) 'Social structure and biased technical change in Argentine agriculture', in Binswanger and Ruttan (eds) (1978), pp.297–323.

DEMSETZ, H. (1967) 'Toward a theory of property rights', *American Economic Review*, 57(2) pp.347–59, (papers and proceedings).

DENISON, E.F. (1962) *Sources of Economic Growth in the United States and the Alternatives Before Us*, New York: Committee for Economic Development.

DE SILVA, N.T.M.H., and TISDELL, C.A. (1985) 'Density related yield functions for coconut (*cocos nucifera*): An empirical estimation procedure', *Experimental Agriculture*, 21, pp.259–69.

DIWAN, R.K. and KALLIANPUR, R. (1985) 'Biological technology and land productivity: Fertilizers and food production in India', *World Development*, 13(5), pp.627–38.

—— (1986) *Productivity and Technical Change in Foodgrains*, New Delhi: Tata McGraw-Hill.

DIXON, W.J. (1983) *BMDP Statistical Software*, Los Angeles: University of California Press.

DOUGLASS, G.K. (1984) 'The meanings of agricultural sustainability', in G.K. Douglass (ed.) *Agricultural Sustainability in a Changing World Order*, Boulder, Colorado: Westview Press, pp.3–29.

DUNCAN, R. and TISDELL, C.A. (1971) 'Research and technical progress: The returns to producers', *Economic Record*, 47(1), pp.124–9.

EIU (1976) *Quarterly Economic Review of Pakistan, Bangladesh and Afghanistan, Annual Supplement*, London: Economist Intelligence Unit.

EPBS (1969) *Statistical Digest of East Pakistan*, no. 6, Dacca: East Pakistan Bureau of Statistics.

EVANS, L.T. (1986) 'Yield variability in cereals: Concluding assessment', in Hazell (ed.) (1986), pp.1–12.

EVENSON, R.E. and KISLEV, Y. (1973) 'Research and productivity in wheat and maize', *Journal of Political Economy*, 81(6), pp.1309–29.

EVENSON, R.E., O'TOOLE, J.C., HERDT, R.W., COFFMAN, W.R., and KAUFMAN, H.E. (1979) 'Risk and uncertainty as factors of crop

306

Bibliography

improvement research', in J.A. Roumasset, J.M. Boussard and I. Singh (eds), *Risk, Uncertainty and Agricultural Development*, Laguna, Philippines: Southeast Asian Regional Center for Graduate Study and Research in Agriculture and the Agricultural Development Council, pp.249–64.

FAALAND, J. and PARKINSON, J.R. (1976) *Bangladesh: The Test Case of Development*, London: C. Hurst.

FAO (1984) *Land, Food and People*, Rome: Food and Agricultural Organization.

FARMER, B.H. (1979) 'The "Green Revolution" in South Asian ricefields: Environment and production', *Journal of Development Studies*, 15(4), pp.304–19.

FEDER, G. (1980) 'Farm size, risk aversion and the adoption of new technology under uncertainty', *Oxford Economic Papers*, 32(2), pp.263–82.

—— (1982) 'Adoption of interrelated innovations: Complementarity and the impacts of risk, scale, and credit', *American Journal of Agricultural Economics*, 64(1), pp.94–101.

FEDER, G., JUST, R.E. and ZILBERMAN, D. (1985) 'Adoption of agricultural innovations in developing countries: A survey', *Economic Development and Cultural Change*, 33(2), pp.255–98.

FEDER, G. and O'MARA, G.T. (1981) 'Farm size and the diffusion of Green Revolution technology', *Economic Development and Cultural Change*, 30(1), pp.59–76.

—— (1982) 'On information and innovation diffusion', *American Journal of Agricultural Economics*, 64(1), pp.145–7.

FEDER, G. and SLADE, R. (1984) 'The acquisition of information and the adoption of new technology', *American Journal of Agricultural Economics*, 66(3), pp.312–20.

FINLAY, K.W. and WILKINSON, G.N. (1963) 'The analysis of adaptation in a plant breeding programme', *Australian Journal of Agricultural Research*, 14, pp.742–54.

FINNEY, D.J. (1971) *Probit Analysis*, Cambridge: Cambridge University Press.

GHOSE, A.K. (1979) 'Farm size and land productivity in Indian agriculture: A reappraisal', *Journal of Development Studies*, 16(1), pp.27–49.

GILL, G.J. (1981) 'Is there a "Draught Power Constraint" on Bangladesh agriculture?', *Bangladesh Development Studies*, 9(3), pp.1–20.

—— (1983) 'Agricultural research in Bangladesh: Costs and returns', Dhaka: Bangladesh Agricultural Research Council (mimeo).

GOLDFELD, S.M. and QUANDT, R.E. (1972) *Nonlinear Methods in Econometrics*, Amsterdam: North Holland Publishing Company.

GOVERNMENT OF INDIA (1982) *Economic Survey 1981–82*, New Delhi: Government of India, Ministry of Finance.

GRABOWSKI, R. and SIVAN, D. (1983) 'The direction of technological change in Japanese agriculture', *Developing Economies*, 21(3), pp.234–43.

GREENSHIELDS, B. and BELLAMY, M. (eds., 1989) *Government Intervention in Agriculture: Cause and Effect*, Aldershot, Hants: Dartmouth Publishing Company Limited.

GRILICHES, Z. (1957) 'Hybrid corn: An exploration in the economics of agricultural research', *Econometrica*, 25(4), pp.501–22.

HAMBURG, M. (1970), *Statistical Analysis for Decision Making*, New York: Harcourt, Brace and World.

HAMID, M.A., SAHA, S.K., RAHMAN, M.A. and KHAN, A.J. (1978) *Irrigation Technologies in Bangladesh: A Study in Some Selected Areas*, Department of Economics: Rajshahi University, Bangladesh.

HAYAMI, Y. and HERDT, R.W. (1977) 'Market price effects of technological change on income distribution in semi-subsistence agriculture', *American Journal of Agricultural Economics*, 59(2), pp.245–56.

HAYAMI, Y. and RUTTAN, V.W. (1971) *Agricultural Development: An International Perspective*, Baltimore, Md.: Johns Hopkins University Press.

—— (1984) 'The Green Revolution: Inducement and distribution', *Pakistan Development Review*, 23(1), pp.37–63.

—— (1985) *Agricultural Development: An International Perspective*, Baltimore, Md.: Johns Hopkins University Press.

HAZELL, P.B.R. (1982) *Instability in Indian Foodgrain Production*, Research Report no. 30, Washington, D.C.: International Food Policy Research Institute.

—— (1984) 'Sources of increased instability in Indian and U.S. cereal production', *American Journal of Agricultural Economics*, 66(3), pp.302–11.

—— (1985) 'Sources of increased variability in world cereal production since the 1960s', *Journal of Agricultural Economics*, 36(2), pp.145–59.

—— (1986) 'Introduction' in Hazell (ed.) (1986), pp.15–46.

—— (ed.) (1986) *Summary Proceedings of a Workshop on Cereal Yield Variability*, Washington, D.C.: International Food Policy Research Institute.

HERDT, R.W. (1987) 'A retrospective view of technological and other changes in Philippine rice farming, 1965–1982', *Economic Development and Cultural Change*, 35(2), pp.329–49.

HERDT, R.W. and CAPULE, C. (1983) *Adoption, Spread and Production Impact of Modern Rice Varieties in Asia*, Los Baños, Philippines: International Rice Research Institute.

HICKS, J.R. (1940) 'The valuation of social income', *Economica*, 7(2), pp.105–24.

—— (1946) *Value and Capital*, London: Oxford University Press.

—— (1948) 'The valuation of social income – A comment on Professor Kuznets' Reflections', *Economica*, 15(3), pp.163–72.

—— (1980) *Causality in Economics*, Canberra: Australian National University Press.

HOSSAIN, MAHABUB (1977) 'Farm size, tenancy and land productivity: An analysis of farm-level data in Bangladesh agriculture', *Bangladesh Development Studies*, 5(3), pp.285–348.

—— (1978) 'Factors affecting tenancy: The case of Bangladesh agriculture', *Bangladesh Development Studies*, 6(2), pp.139–62.

—— (1979) 'Performance of crop production sector in Bangladesh 1965–78', paper presented at the Bangladesh Economic Association Conference, Dhaka, May.

—— (1980) 'Foodgrain production in Bangladesh: Performance, potential and constraints', *Bangladesh Development Studies*, 8(1–2), pp.39–70.

—— (1984) 'Agricultural development in Bangladesh: A historical perspective', *Bangladesh Development Studies*, 12(4), pp.29–57.

—— (1985) 'Price response of fertilizer demand in Bangladesh', *Bangladesh Development Studies*, 13(3–4), pp.41–66.

—— (1988) *Nature and Impact of the Green Revolution in Bangladesh*, Research Report no. 67, Washington, D.C.: International Food Policy Research Institute.

HOSSAIN, MOSHARAFF (1987) *The Assault that Failed: A Profile of Poverty in Six Villages of Bangladesh*, Geneva: United Nations Research Institute for Social Development.

HOSSAIN, M.M., AHMED, A.U., CHURCH, P.E. and TOWHEED, S.M. (1981) 'Transplanted *aman* paddy cultivation practices: Costs and returns in Thakurgaon Tubewell Project area', Joydevpur: Bangladesh Rice Research Institute.

HOSSAIN, M.M. and HARUN, M.E. (1983) 'Costs and returns for transplant *aman* cultivation in Joydebpur 1981–82', Joydevpur: Bangladesh Rice Research Institute.

HOSSAIN, M.M., HARUN, M.E., CHURCH, P.E., AHMED, A.U. and HUQ, Z. (1982) 'Costs and returns for transplanted aman cultivation in Rajshahi, 1980–81', Joydevpur: Bangladesh Rice Research Institute.

HUCK, S.W., CORMIER, W.H. and BOUNDS, W.G. (1974) *Reading Statistics and Research*, New York: Harper and Row.

HUQ, M. (1963) *The Strategy of Economic Planning*, Karachi: Oxford University Press.

IFDC (1982) *Agricultural Production, Fertilizer Use, and Equity Considerations, Results and Analysis of Farm Survey Data, 1979–80, Bangladesh*, Muscle Shoals, Alabama: International Fertilizer Development Center.

IRRI (1974) *Research Highlights for 1974*, Los Baños: International Rice Research Institute.

—— (1978) *Interpretive Analysis of Selected Papers from Changes in Rice Farming in Selected Areas of Asia*, Los Baños, Philippines: International Rice Research Institute.

ISLAM, N. (1974) 'The state and prospects of the Bangladesh economy', in E.A.G. Robinson and K.B. Griffin (eds), *The Economic Development of Bangladesh within a Socialist Framework*, London: Macmillan, pp.1–15.

ISLAM, R. (1979) 'What has been happening to rural income distribution in Bangladesh?', *Development and Change*, 10(3), pp.385–402.

—— (1985) 'Non-farm employment in Asia: Issues and evidence', in Shand (ed.) (1985) vol. 1, pp.153–74.

IUCN (1980) *World Conservation Strategy: Living Resources Conservation for Sustainable Development*, Glands, Switzerland: International Union for the Conservation of Nature and Natural Resources.

JABBAR, M.A. (1977) 'Relative productive efficiency of different tenure classes in selected areas of Bangladesh', *Bangladesh Development Studies*, 5(1), pp.17–50.

JAIN, H.K., DAGG, M. and TAYLOR, T.A. (1986) 'Yield variability and the transition of the new technology', in Hazell (ed.) (1986), pp.77–85.

JANUZZI, F.T. and PEACH, J.T. (1980) *The Agrarian Structure of Ban-*

gladesh: An Impediment to Development, Boulder, Colorado: Westview Press.

JODHA, N.S. (1985) 'Population growth and the decline of common property resources in Rajasthan, India', *Population and Development Review*, 11(2), pp.247–64.

—— (1986) 'Common property resources and rural poor in dry regions of India', *Economic and Political Weekly*, 21(27), pp.1169–81.

JOHANSEN, L. (1972) *Production Functions*, Amsterdam: North-Holland Publishing Company.

JOHNSTON, B.F. and MELLOR, J.W. (1961) 'The role of agriculture in economic development', *American Economic Review*, 51(4), pp.566–93.

JOHNSTON, J. (1984) *Econometric Methods*, New York: McGraw-Hill Book Company.

JONES, S. (1984) 'Agrarian structure and agricultural innovations in Bangladesh: Panimara village, Dhaka district', in T.P. Balyss-Smith and S. Wanmali (eds) (1984), pp.194–211.

JOSHI, P.K. and KANEDA, H. (1982) 'Variability in yield in foodgrain production', *Economic and Political Weekly*, 17(13), pp.A2–8.

KANEDA, H. (1982) 'Specification of production functions for analyzing technical change and factor inputs in agricultural development', *Journal of Development Economics*, 11(1), pp.97–108.

KAUTSKY, K. (1899) *Die Agrafare*, Stuttgart.

KENDALL, M.G. (1980) *Multivariate Analysis*, London: Charles Griffin.

KENDRICK, J.W. (1956) 'Productivity trends, capital and labour', *Review of Economics and Statistics*, 38(3), pp.248–57.

—— (1961) *Productivity Trends in the United States*, Princeton, N.J.: Princeton University Press.

KHAN, A.R. (1972) *The Economy of Bangladesh*, London: Macmillan.

—— (1979) 'The Comilla model and the integrated rural development programme of Bangladesh: An experiment in cooperative capitalism', *World Development*, 7(4–5), pp.397–422.

—— (1984) 'Real wages of agricultural workers in Bangladesh', in A.R. Khan and E. Lee (eds), *Poverty in Rural Asia*, Bangkok: Asian Employment Programme (ARTEP), International Labour Office.

KHAN, M. (1981) 'Technical efficiency and fertilizer use in Bangladesh agriculture', in Mahmud (ed.) (1981), pp.142–61.

KISLEV, Y. and SCHCHORI-BACHRACH, N. (1973) 'The process of an innovation cycle', *American Journal of Agricultural Economics*, 55(1), pp.28–37.

KMENTA, J. (1971) *Elements of Econometrics*, New York: Macmillan.

KOUTSOYIANNIS, A. (1978) *The Theory of Econometrics*, London: Macmillan.

KRUTILLA, J.V. (1967) 'Conservation reconsidered', *American Economic Review*, 57, pp.777–86.

KRUTILLA, J.V. and FISHER, A.C. (1975) *The Economics of Natural Environments*, Washington, D.C.: Resources for the Future.

KUZNETS, S. (1948) 'On the valuation of social income – Some reflections on Professor Hicks' article', *Economica*, 15(1), pp.1–16 and 15(2), pp.116–31.

—— (1963) 'Quantitative aspects of the economic growth of nations: VIII',

Economic Development and Cultural Change, 11(1), pp.1–80.

LEABO, D.A. (1972) *Basic Statistics*, Homewood, Ill.: Richard D. Irwin.

LINDNER, R., FISCHER, A. and PARDEY, P. (1979) 'The time to adoption', *Economic Letters*, 2, pp.187–90.

LINDNER, R.K. and JARRETT, F.G. (1978) 'Supply shifts and the size of research benefits', *American Journal of Agricultural Economics*, 60(1), pp.48–58.

LIPTON, M. (1978) 'Inter-farm, inter-regional and farm–non-farm income distribution: The impact of new cereal varieties', *World Development*, 6(3), pp.319–37.

—— (1985) *Land Assets and Rural Poverty*, World Bank Staff Working Paper no. 744, Washington, D.C.: World Bank.

LITTLE, I.M.D. (1960) *A Critique of Welfare Economics*, London: Oxford University Press.

LUND, P.J., IRVING, R.W. and CHAPMAN, W.G. (1980) 'The economic evaluation of publicly-funded agricultural research and development', *Oxford Agrarian Studies*, 9, pp.14–33.

MAHMUD, W. (1979) 'Foodgrains demand elasticities of rural households in Bangladesh: An analysis of pooled cross-section data', *Bangladesh Development Studies*, 7(1), pp.59–70.

—— (ed.) (1981) *Development Issues in an Agrarian Economy – Bangladesh*, Dhaka: Centre for Administrative Studies.

MAITRA, P. (1980) *The Mainspring of Economic Development*, London: Croom Helm.

—— (1988) 'Population growth, technological change and economic development: The Indian case, with a critique of Marxist interpretation', in Tisdell and Maitra (eds) (1988), pp.9–33.

MALTHUS, T.R. (1798) *An Essay on the Principles of Population*, London: Macmillan, (Reprinted for the Royal Economic Society, 1926)

MANSFIELD, E.S. (1961) 'Technical change and the rate of imitation', *Econometrica*, 29(4), pp.741–66.

—— (1968) *Industrial Research and Technological Innovation*, New York: W.W. Norton.

MARKOWITZ, H.M. (1959) *Portfolio Selection – Efficient Diversification of Investments*, Cowles Foundation Monograph no. 16, New York: John Wiley.

MARSHALL, A. (1890) *Principles of Economics*, 1st edn, London: Macmillan.

MEHRA, S. (1981) *Instability in Indian Agriculture in the Context of the New Technology*, Research Report no. 25, Washington, D.C.: International Food Policy Research Institute.

MELLOR, J.W. (1976) *The New Economics of Growth*, Ithaca: Cornell University Press.

—— (1978) 'Food price policy and income distribution in low income countries', *Economic Development and Cultural Change*, 27(1), pp.1–26.

—— (1985) 'Determinants of rural poverty: The dynamics of production and prices', in Mellor and Desai (eds) (1985), pp.21–40.

MELLOR, J.W. and DESAI, G.M. (eds) (1985), *Agricultural Change and*

Rural Poverty: Variations on a Theme by Dharm Narain, Baltimore, Md.: Johns Hopkins University.

MINHAS, B.S. and VAIDYANATHAN, A. (1965) 'Growth of crop output in India, 1951–54 to 1958–61: An analysis by component elements', *Journal of the Indian Society of Agricultural Statistics*, 17(2), pp.230–52.

MORAWETZ, D. (1974) 'Employment implications of industrialisation in developing countries: A survey', *Economic Journal*, 84(3), pp.492–542.

MURSHID, K.A.S. (1986) 'Instability in foodgrain production in Bangladesh: Nature, levels and trends', *Bangladesh Development Studies*, 14(2), pp.33–68.

MUTHIA, C. (1971) 'The Green Revolution: Participation by large versus small farmers', *Indian Journal of Agricultural Economics*, 26(1), pp.53–66.

MYRDAL, G. (1968) *Asian Drama*, New York: Twentieth Century Fund.

NARAIN, D. (1977) 'Growth of productivity in Indian agriculture', *Indian Journal of Agricultural Economics*, 32(1), pp.1–44.

OLDFIELD, M. (1989) *The Value of Conserving Genetic Resources*, Sunderland, Mass.: Sinaver Associates.

OZGA, S.A. (1960) 'Imperfect markets through lack of knowledge', *Quarterly Journal of Economics*, 74(1), pp.19–52.

PALMER, I. (1976) *The New Rice in Asia: Conclusions from Four Country Studies*, Geneva: United Nations Research Institute of Social Development.

PAPANEK, G.F. (1967) *Pakistan's Development: Social Goals and Private Incentives*, Cambridge, Mass.: Harvard University Press.

PARTHASARATHY, G. (1984) 'Growth rates and fluctuations of agricultural production: A district-wide analysis in Andhra Pradesh', *Economic and Political Weekly*, 19(26), pp.A74–84.

PARTHASARATHY, G. and PRASAD, D.S. (1978) 'Response to the impact of the new rice technology by farm size and tenure: Andhra Pradesh, India', in IRRI (1978), pp.111–27.

PEARSE, A. (1980) *Seeds of Plenty, Seeds of Want*, Oxford: Clarendon Press.

PLUCKNETT, D.P., SMITH, N.J.H., WILLIAMS, J.J. and ANISHETY, N.M. (1986) *Gene Bank and the World's Food Supply*, Princeton, N.J.: Princeton University Press.

POIRIER, D.J. (1976) *The Econometrics of Structural Change*, Amsterdam: North Holland Publishing Company.

POSNER, R. (1980) 'A theory of primitive society with special reference to primitive law', *Journal of Law and Economics*, 23(1), pp.1–55.

PRAY, C.E. (1979) 'The economics of agricultural research in Bangladesh', *Bangladesh Journal of Agricultural Economics*, 2(2), pp.1–34.

—— (1980) 'An assessment of the accuracy of the official agricultural statistics of Bangladesh', *Bangladesh Development Studies*, 8(1–2), pp.1–38.

PRAY, C.E. and ANDERSON, J.R. (1985) *Bangladesh and the CGIAR Centers: A Study of their Collaboration in Agricultural Research*, CGIAR Study Paper no. 8, Washington, D.C.: World Bank.

QUASEM, M.A. (1986) 'The impact of privatisation on entrepreneurial development in Bangladesh agriculture', *Bangladesh Development Studies*, 14(2), pp.1–20.

QUIGGIN, J. (1982) 'A theory of anticipated utility', *Journal of Economic Behaviour and Organization*, 3, pp.323–43.

RAHIM, M.A. (1977) '*Aman* crop survey in Comilla Kotwali Thana (1975)', Comilla: Bangladesh Academy for Rural Development.

RAHMAN, ATIQUR (1981) 'Adoption of new technology in Bangladesh: Testing some hypotheses', in Mahmud (ed.) (1981), pp.55–77.

RAHMAN, ATIUR (1982) 'Land concentration and dispossession in two villages of Bangladesh', *Bangladesh Development Studies*, 10(2), pp.51–84.

RAHMAN, R.I. (1983) 'Adoption of HYV: Role of availability and the supply-side problems', *Bangladesh Development Studies*, 11(4), pp.61–75.

RANDALL, A. (1986) 'Human preferences, economics and preferences and the preservation of species', in B.G. Norton (ed.), *The Preservation of Species: The Value of Biological Diversity*, Princeton, N.J.: Princeton University Press.

RAO, C.H.H. (1975) *Technological Change and Distribution of Gains in Indian Agriculture*, Delhi: Macmillan.

RAY, S.K. (1983) 'An empirical investigation into the nature and causes for growth and instability in Indian agriculture: 1950–1980', *Indian Journal of Agricultural Economics*, 38(4), pp.459–74.

REDCLIFT, M. (1987) *Sustainable Development: Exploring the Contradictions*, London: Methuen.

REDDY, V.N. (1978) 'Growth rates', *Economic and Political Weekly*, 13(19), pp.806–12.

REMENYI, J.V. (1988) 'Partnership in research: A new model for development assistance', in Tisdell and Maitra (eds) (1988), pp.91–113.

REPETTO, R. and HOLMES, T. (1983) 'The role of population in resource depletion in developing countries', *Population and Development Review*, 9(4), pp.609–32.

RICARDO, D. (1817) *Principles of Political Economy and Taxation*, New York: E.P. Dutton (Everyman's edition, 1911).

ROGERS, E.M. and SHOEMAKER, F.F. (1971) *Communication of Innovations: A Cross-Cultural Approach*, New York: Free Press.

ROY, A.D. (1952) 'Safety first and holding of assets', *Econometrica*, 20(3), pp.431–49.

RUDRA, R. (1970) 'The rate of growth of the Indian economy', in E.A.G. Robinson and M. Kidron (eds), *Economic Development in South Asia*, London: Macmillan, pp.35–53.

RUNGE, C.F. (1981) 'Common property externalities: Isolation, assurance, and resource depletion in a traditional grazing context', *American Journal of Agricultural Economics*, 63(4), pp.595–606.

RUTTAN, V.W. (1977) 'The Green Revolution: Seven generalizations', *International Development Review*, 19(4), pp.16–23.

—— (1987) 'Agricultural scientists as reluctant revolutionaries', *Choices*, 3rd Quarter, p.3.

—— (1988) 'Sustainability is not enough', *Alternative Agriculture*, 3(2–3), pp.128–30.

RUTTAN, V.W., BINSWANGER, H.P., HAYAMI, Y., WADE, W.W. and WEBER, A. (1978) 'Factor productivity and growth: A historical interpretation', in Binswanger and Ruttan (eds) (1978), pp.44–87.

SAGAR, V. (1977) 'A component analysis of the growth of productivity and production in Rajasthan: 1956–61 to 1969–74', *Indian Journal of Agricultural Economics*, 32(1), pp.108–19.

—— (1980) 'Decomposition of growth trends and certain related issues', *Indian Journal of Agricultural Economics*, 35(2), pp.42–69.

SAHA, B.K. (1978) *Socio-Economic Effects of Technological Change in Bangladesh Agriculture: A Study of Two Villages in Bangladesh*, unpublished M. Phil thesis, Rajshahi: Rajshahi University, Bangladesh.

SAMUELSON, P.A. (1948) *Foundations of Economic Analysis*, Cambridge, Mass.: Harvard University Press.

SARMA, J.S. and ROY, S. (1979) *Two Analyses of Indian Foodgrain Production and Consumption Data*, Research Report no. 12, Washington D.C.: International Food Policy Research Institute.

SAWANT, S.D. (1983) 'Investigation of the hypothesis of deceleration in Indian agriculture', *Indian Journal of Agricultural Economics*, 38(4), pp.475–96.

SCHLUTER, M. (1971) *Differential Rates of Adoption of New Seed Varieties in India: The Problem of the Small Farm*, US Agency for International Development, Occasional Paper no. 47, Ithaca, New York: Cornell University.

SCHULTZ, T.W. (1964) *Transforming Traditional Agriculture*, New Haven: Yale University Press.

—— (1974) 'Is modern agriculture consistent with a stable environment?' in *The Future of Agriculture: Technology, Policies and Adjustment*, papers and reports, 15th International Conference of Agricultural Economists, Oxford: Oxford Agricultural Economics Institute, pp.235–42.

SCHUTJER, W. and VAN DER VEEN, M. (1977) *Economic Constraints on Agricultural Technology Adoption in Developing Countries*, Occasional Paper no. 5, Washington, D.C.: United States Agency for International Development.

SEN, A.K. (1960) *Choice of Techniques*, Oxford: Basil Blackwell.

—— (1966) 'Peasants and dualism with or without surplus labour', *Journal of Political Economy*, 74(5), pp.425–50.

—— (1975) *Employment, Technology and Development*, Oxford: Clarendon Press.

—— (1981) *Poverty and Famines: An Essay on Entitlement and Deprivation*, Oxford: Clarendon Press.

SEN, B. (1974) *The Green Revolution in India: A Perspective*, New Delhi: Wiley Eastern.

SEN, S.R. (1967) 'Growth and instability in Indian agriculture', *Journal of the Indian Society of Agricultural Statistics*, 19(1), pp.1–30.

SHAHABUDDIN, Q., MOSTELMAN, S. and FEENY, D. (1986) 'Peasant behaviour towards risk and socio-economic and structural characteristics of farm households in Bangladesh', *Oxford Economic Papers*, 38(1), pp.122–30.

SHAND, R.T. (ed.) (1985) *Off-Farm Employment in the Development of Rural Asia*, Canberra: National Centre for Development Studies, Australian National University (2 vols).

SHARMA, A.C. (1973) 'Influence of certain economic and technological factors on the distribution of cropped area under various crops in the Ludhiana district', *Journal of Research: Punjab Agricultural University*, 10, pp.243–9.

SHETTY, N.S. (1968) 'Agricultural innovations: Leaders and laggards', *Economic and Political Weekly*, 3(33), pp.1271–82.

SINHA, S. (1984) 'Growth of scientific temper: Rural context', in M. Gibbons, P. Gummett, and B. Udgaonkar (eds) *Science and Technology Policy in the 1980s and Beyond*, London: Longman, pp.166–90.

SMITH, V.K. and KRUTILLA, J.V. (1979) 'Endangered species, irreversibility and uncertainty: A comment', *American Journal of Agricultural Economics*, 61, pp.371–5.

SOLOW, R.M. (1957) 'Technical change and the aggregate production function', *Review of Economics and Statistics*, 39(3), pp.312–20.

STAUB, W.J. and BLASE, M.G. (1974) 'Induced technical change in developing agricultures: Implication for income distribution and development', *Journal of Developing Areas*, 8(4), pp.581–95.

TAYLOR, C.R. (1980) 'The nature of benefits of costs of use of pest control methods', *American Journal of Agricultural Economics*, 62(5), pp.1007–11.

THEIL, H. (1971) *Principles of Econometrics*, New York: John Wiley.

THIBODEAU, R. and FIELD, H. (1984) *Sustaining Tomorrow: A Strategy for World Conservation and Development*, Hanover: University Press of New England.

THOMAS, G.B. (1968) *Calculus and Analytic Geometry, Part Two*, London: Addison-Wesley.

THORNER, D., KERBLAY, B. and SMITH, R.E.F. (eds) (1966) *A.V. Chayanov on the Theory of Peasant Economy*, Homewood, Ill.: Richard D. Irwin.

TINTNER, G. (1965) *Econometrics*, New York: John Wiley.

TISDELL, C.A. (1962) 'Decision making and the probability of loss', *Australian Economic Papers*, 1(1), pp.109–18.

—— (1972) *Microeconomics: The Theory of Economic Allocation*, Sydney: John Wiley.

—— (1983a) 'Conserving living resources in the Third World countries: Economic and social issues', *International Journal of Environmental Studies*, 22, pp.11–24.

—— (1983b) 'The optimal choice of a variety of a species for variable environmental conditions', *Journal of Agricultural Economics*, 34(2), pp.175–85.

—— (1983c) 'An economist's critique of the world conservation strategy with some examples from Australian experience', *Environmental Conservation*, 10, pp.43–52.

—— (1984) 'The provision of wilderness by clubs', *Revista Internazionale Di Scienze Economiche E Commerciali*, 31, pp.758–61.

—— (1985a) 'The world conservation strategy, economic policies and sustainable resource-use in developing countries', *Environmental Professional*, 7, pp.102–7.

—— (1985b) 'Economics, ecology, sustainable agricultural systems and development', *Development Southern Africa*, 2, pp.512–21.

—— (1988) 'Sustainable development: Differing perspectives of ecologists and economists, and relevance to LDCs', *World Development*, 16(3), pp.373–84.

—— (1990a) 'Economics and the debate about preservation of species, crop varieties and genetic diversity', *Ecological Economics* 2, pp.77–90.

—— (1990b) *Natural Resources, Growth and Development: Economics, Ecology and Resource-Scarcity*, New York: Praeger.

TISDELL, C.A. and ALAUDDIN, M. (1988a) 'Transferring and developing agricultural technology for LDCs: Multinational corporations and other means', *Asian Journal of Economic and Social Studies*, 7(1), pp.24–36.

—— (1988b) 'Diversification and stability implications of new crop varieties': Theoretical and empirical evidence', a paper presented at the 32nd Annual Conference of the Australian Agricultural Economics Society, La Trobe University, Bundoora, Victoria 3083, February 9–11.

—— (1989) 'New crop varieties: Impact on diversification and stability of yields', *Australian Economic Papers*, pp.123–40.

TISDELL, C.A. and FAIRBAIRN, I.J. (1984) 'Subsistence economies, and unsustainable development and trade: Some simple theory', *Journal of Development Studies*, 20(2), pp.227–41.

TISDELL, C.A. and MAITRA, P. (eds) (1988) *Technological Change, Development and the Environment: Socio-Economic Perspectives*, London: Routledge and Kegan Paul.

TURNER, R.K. (1988) *Sustainable Environmental Management: Principles and Practice*, London: Belhaven Press.

USAID (1982) *Joint Bangladesh and U.S. Government Evaluation of the Fertilizer Distribution Improvement Project (388–0024)*, Dhaka: Agriculture and Forest Division, Bangladesh Ministry of Agriculture.

VENEGAS, E.C. and RUTTAN, V.M. (1964) 'An analysis of rice production in the Philippines', *Economic Research Journal*, 11(1), pp.159–78.

VENKATARAMANAN, L.S. and PRAHLADACHAR, M. (1980) 'Growth rates and cropping pattern changes in agriculture in six states: 1950 to 1975', *Indian Journal of Agricultural Economics*, 35(2), pp.71–84.

WCED (1987) *Our Common Future*, World Commission on Environment and Development, Oxford: Oxford University Press.

WENNERGREN, E.B., ANTHOLT, C.H. and WHITAKER, M.D. (1984) *Agricultural Development in Bangladesh*, Boulder, Colorado: Westview Press.

WHARTON, C.R. (ed.) (1969) *Subsistence Agriculture and Economic Development*, London: Frank Cass.

WISE, W.S. (1978) 'The economic analysis of agricultural research', *R & D Management*, 8(3), pp.185–9.

—— (1981) 'The theory of agricultural research benefits', *Journal of Agricultural Economics*, 32(2), pp.147–57.

WORLD BANK (1982) *Bangladesh: Foodgrain Self-Sufficiency and Crop Diversification (Annexes and Statistical Appendix)*, Report no. 3953-BD.

—— (1986) *World Development Report*, New York and Oxford: Oxford University Press.

WYNEN, E. (1989) *Sustainable and Conventional Agriculture: An Economic Analysis of Cereal-Livestock Farming*, unpublished PhD thesis, LaTrobe University, Australia.

YAMADA, S. and V.W. RUTTAN (1980) 'International comparisons of productivity in agriculture', in J.W. Kendrick and B.N. Vaccara (eds), *New Developments in Productivity Measurement and Analysis*, Chicago: University of Chicago Press, pp.509–85.

YOTOPOULOS, P.A. and LAU, L.J. (1973) 'A test for relative economic efficiency: Some further results', *American Economic Review*, 63(1), pp.214–23.

YOTOPOULOS, P.A. and NUGENT, J.B. (1976) *Economic Development: Empirical Investigations*, New York: Harper and Row.

YUDELMAN, M., BANERJI, R. and BUTLER, G. (1970) 'The use of an identity to examine the association between technological changes and aggregate labour utilization in agriculture', *Journal of Development Studies*, 7(1) pp.36–49.

Index